AGRICULTURE ISSUES AND POLICIES

CASSAVA

FARMING, USES, AND ECONOMIC IMPACT

AGRICULTURE ISSUES
AND POLICIES

Additional books in this series can be found on Nova's website
under the Series tab.

Additional E-books in this series can be found on Nova's website
under the E-books tab.

AGRICULTURE ISSUES AND POLICIES

CASSAVA

FARMING, USES, AND ECONOMIC IMPACT

COLLEEN M. PACE
EDITOR

Nova Science Publishers, Inc.
New York

NOTICE TO THE READER

Library of Congress Cataloging-in-Publication Data

ISBN 978-1-61209-655-1
Cassava : farming, uses, and economic impact / editor: Colleen M. Pace.
    p. cm.
  Includes index.
  ISBN 978-1-61209-655-1 (hardcover)
  1. Cassava. 2. Cassava--Utilization. 3. Cassava--Industrial
applications. 4. Cassava--Economic aspects. I. Pace, Colleen M.
  SB211.C3C382 2011
  633.6'82--dc22
                                                                    2011001142

Published by Nova Science Publishers, Inc. ✛ New York

CONTENTS

PREFACE

Cassava, also known as tapioca or manioc, is one of the major root crops in more than 100 countries of the humid tropics and sub-tropics. This new book presents topical research in the study of the farming, uses and economic impact of cassava. Topics discussed include potential uses of cassava wastewater in biotechnological processes; cassava starch and flour in the production of bio-ethanol, bio-plastics, acetone-butanol, dextrin, sugar syrups and organic acids and cassava starch as a biodegradable polymer material. (Imprint: Nova Press)

Chapter 1 - Cassava is one of the richest source of starch. The tuberous roots contain up to 35% of starch and are low in proteins, soluble carbohydrates and fats that make starch extraction from cassava comparatively easier. Cassava starch and flour serve as raw materials for production of a number of industrial fermented products such as bio-ethanol, bio-plastics, acetone-butanol, dextrin, sugar (glucose and fructose) syrups and organic (lactic and glutamic) acids, microbial polysaccharides, baker's yeasts, etc. Bio-ethanol is produced from cassava starch and flour by liquefaction and saccharification, and subsequent fermentation by microorganisms. Cassava starch is converted to lactic acid by lactic acid bacteria and polymerization of lactic acid by thermochemical reactions lead to the production of polylactic acid, which is then blended with other polymers to yield biodegradable plastics. In this chapter, the recent developments in the technology in the production of the above bio-products from cassava starch and flour have been discussed.

Chapter 2 - Cassava flour is the main cassava derivative for food use in Brazil and its processing generates solid and liquid residues. Concerning the latter, the wash water plus the water extracted from the roots by squeezing is denominated *cassava wastewater* or *manipueira*, and is mainly composed of nutrients such as nitrogen, carbon, potassium, phosphorus, zinc, manganese, calcium, magnesium, sulfur, copper, iron and sodium. This residue is very harmful to the environment due to its high BOD and cyanogenic glycosides. However, the use of this by-product is evident in many areas as an alternative low-cost carbon substrate for use in high-value market compounds. The use of agroindustrial residues in biotechnological processes has been indicated as an approach to reduce the volume of waste released directly into the environment or involving high costs for effluent treatment. This chapter discusses some applications of this residue in the production of some enzymes (*i.g.* amylases), citric acid, aroma compounds (*i.g.* alcohols), biotransformation processes (*i.g.* culture medium for terpene biotransformation), production of ethanol and, principally, the production of biosurfactants.

Chapter 3 - Cassava is a staple food to millions of people in tropical and subtropical countries. Although it is traditionally cultivated from stem cuttings, which is a simple and inexpensive technique, this method presents serious problems such as low multiplication rates, difficulties to conserve stems, and dissemination of pests and diseases. Many of these problems would be solved through in vitro tissue culture. This chapter evaluates the in vitro establishment and multiplication of 28 cassava clones of agronomic interest for the Northeastern Argentina, a boundary area for this crop. Since the transfer of in vitro plants to ex vitro conditions is a critical phase of micropropagation, the effect of different acclimatization treatments are evaluated on survival and growth parameters of plants (cv EC118) grown in a culture chamber. Also shown are their scored field survival and performance by comparing them with plants obtained by the conventional planting technique. After disinfection, uninodal segment culture in Murashige and Skoog medium supplemented with 0.01 mg/L BAP + 0.01 mg/L NAA + 0.1 mg/L GA3 allowed the in vitro establishment of 100% of the clones and their subsequent multiplication. Cultures were maintained at 27°±2°C with a 14 h photoperiod. During establishment, sprouting occurred in 100% of the clones and rooting in 93% of them; the remaining clones formed roots during the multiplication phase. Thirty days after multiplication, the plants presented significant differences in plant height, average number of nodes per plant and number of roots per plant. During acclimatization, five treatments were evaluated: three substrates (perlite, T1; sand + vermicompost, T2; commercial substrate composed of peat and perlite, T3), and two hydroponic treatments (tapwater, T4; Arnon and Hoagland nutrient solution, T5). Although in chamber growth conditions the acclimatized plants showed statistical differences in several growth parameters depending on the treatments, no differences were observed in the survival percentage. Shoot and root fresh and dry weight and leaf area were highest in T5 and lowest in T2 and T4. Field survival differed significantly between treatments, discriminating a group with high survival rates (T5: 73.3%, T3: 86.7%, and control treatment: 100%) and another with low survival rates (T2: 33.3%; T1: 35% and T4: 36.7%). At harvest, there were no significant differences in the total fresh weight. However, the percentage of biomass partitioned to roots was significantly higher in T3 and T5, which resulted in a higher tuberous roots yield than that of the control treatment.

Chapter 4 - Advances in industrial biotechnology offer potential opportunities for the economic utilization of agro-industrial residues, such as those from the cassava processing industry. Three main types of residue are generated during the industrial processing of cassava: peels and bagasse (solid); and wastewater (liquid). Both types of waste are poor in protein content, but are carbohydrate-rich residues and generated in large amounts during the production of flour (which generates more solid residues) and starch (which generates more liquid residues) from the tubers. Waste treatment and disposal costs constitute a huge financial burden to the cassava processing industry as well as an environmental problem. Therefore, there is a great need for the better management of these waste products. Due to its rich organic nature, cassava residue can serve as an ideal substrate for microbial processes in the production of different products. Attempts have been made to produce products such as organic acids, flavor and aroma compounds, mushrooms, methane and hydrogen gas, enzymes, ethanol, lactic acid, biosurfactants, polyhydroxyalkanoate, essential oils, xanthan gum and fertilizer from cassava bagasse, peels and wastewater. The use of cassava residues as feedstock in biotechnological processes is a viable alternative that can contribute toward a reduction in production costs, an increase in the economic value of these residues and the minimization of environmental problems related to waste discharge. This study reviews processes and products developed for aggregating value to cassava residues through biotechnological means, demonstrating the potential of this agro-industrial raw material.

Chapter 5 - Cassava is produced in Latin America, Asia and Southern Africa. In Latin America, cassava is popularly used as a meal, as animal fodder or cooked and eaten as a vegetable and part of its production is exported. It has been seen that cassava starch is used to a much lesser extent than other starches, like corn one, in food industry. Anyhow, its importance as a source of starch is growing rapidly, especially because its price in the world market is low when compared to starches from other sources.

Edible films and coatings are not designed for totally replacing traditional packaging but to help, as an additional stress factor, for protecting food products, improving quality and shelf life without impairing consumer acceptability. They can control moisture, gases, lipid migration and can also be carriers of additives and nutrients. Cellulose, gums, starch and proteins have been used to formulate edible films and plasticizers are usually employed to enhance its mechanical properties.

The objective of this chapter is to analyze the use of cassava starch for edible films and coatings formulation. The barrier and mechanical characteristics of these edibles according to production technique and formulation as well as their effectiveness for supporting different antimicrobials is considered. It is also revised the possibility of their application to food products.

Chapter 6 - Biodegradable polymers have gained great attention of researchers in decades ago. Biodegradable polymers are environment friendly products which are able to reduce environmental pollution problems. In recent decade, one of most important in the development of biodegradable polymer area is to produce cheap starch based biodegradable polymer. Native starch is suitable to produce biodegradable polymer material because it is available abundantly at low cost. Starch is harvested from varieties of crops such as corn, potato, sago, cassava, wheat and etc. Among the crops, cassava is most widely growth to produce sustainable and cheap source of starch globally. In this chapter, cassava starch is utilized to produce biodegradable polymer compound. Cassava starch (CSS) is required to undergo series of processes in order to produce the CSS based polymer products by existing polymer injection moulding technology. CSS is blended with polyvinyl alcohol, glycerol, and processing aids to improve the processability by injection moulding. For instance, the polyvinyl alcohol is incorporated to enhance the physicomechanical properties of the CSS. Meanwhile, glycerol is added to CSS for lubrication and gelatinization purpose. All these ingredients need to be melt compounded by twin screw extruder into pellet form. The PVA-CSS polymer compound can be used to produce plastic articles by using injection moulding technique under optimum processing conditions. The amount of CSS in polymer compound influences the mechanical and thermal properties of the polymer compounds.

Chapter 7 - Cassava, ranking presently as the world's fifth large crop in starch production, is shedding light on its importance in the global agricultural economy due to its various biological characters. It is a perennial crop, easy and economic in cultivation. It requires little or no fertilizers, beneficial for environments. Cassava is economic in land usage. It can utilize low quality land such as semi-dry and mountainous land. Importantly, it can produce large tuberous roots that result in a high starch yield. Moreover its above-ground biomass can be employed as industrial feedstocks. This review will from an economic point of view elucidate the importance and possible contribution of cassava to the world's future agricultural economy. By analyzing the use-value, economic benefits, threats to cassava farming, directions of traditional and molecular breeding, possible added-value to the future cassava and the World production in the past twenty years, the possibility of whether cassava can be used as a multipurpose crop to meet the requirements for future plants is discussed.

Chapter 8 - Tropical ataxic neuropathy (TAN) and konzo are two neurological disorders associated with the chronic consumption of cassava (*Manihot esculenta*) in several African countries. TAN is characterized by sensory polyneuropathy, sensory ataxia, bilateral optic atrophy and bilateral sensori-neural deafness. It occurs in poor, undernourished elderly individuals subsisting on a monotonous cassava diet with minimal protein supple mentation. Konzo is a syndrome of symmetrical, non-remitting, non-progressive spastic paraparesis, with a predilection for children and women of child-bearing age. It is invariably associated with monotonous consumption of inadequately processed bitter cassava roots with very minimal protein supplementation. Chronic cyanide intoxication from consumption of cyanogenic glycosides in cassava was long thought to be the major etiological factor for TAN, but there has been no evidence of a causal association. Similarly, high cyanide consumption with low dietary sulfur intake due to almost exclusive consumption of insufficiently processed bitter cassava roots was proposed as the cause of konzo, but there has also been no evidence of a causal association. The roles of the cyanogenic glycoside linamarin, acetone cyanohydrin and cyanate in the etiology of konzo had been evaluated, but there was no evidence of a causal association. The etiologies of both TAN and konzo therefore remained unknown, despite studies in several countries aimed at unraveling the etiologic mechanisms of these debilitating diseases.

In this chapter an etiological mechanism of thiamine deficiency for both TAN and konzo is discussed. It is postulated that in TAN and konzo patients, thiamine deficiency results from the inactivation of thiamine that occurs when, in the absence of dietary sulfur-containing amino acids in these patients with poor protein intake; the sulfur in thiamine is utilized for the detoxification of cyanide in the cyanogenic glycosides consumed in cassava. Thiamine is known to be rendered inactive when the sulfur in its thiazole moiety is combined with hydrogen cyanide. Evidence from the literature implicating chronic thiamine deficiency in the etiology of TAN are discussed. These include evidence of abnormal pyruvate metabolism reversed by thiamine in patients with TAN, evidence from erythrocyte transketolase activity indicating significant thiamine deficiency in patients with TAN compared to controls, and a placebo-controlled trial of therapeutic doses of thiamine which showed a clinically dramatic and statistically significant improvement in ataxia.

Thiamine status has never been evaluated in patients with konzo, and a therapeutic trial of thiamine has not been conducted. Evidences in support of thiamine deficiency as the etiological mechanism of konzo include a demonstrated evidence of widespread thiamine deficiency in the susceptible population, animal studies demonstrating the similarity of the clinical presentations of konzo to those of thiamine deficiency, and the predilection of konzo for adolescent children and women of child-bearing age; population groups particularly vulnerable to symptomatic thiamine deficiency in the presence of inadequate thiamine intake. Studies are currently underway to confirm the role of thiamine deficiency in the etiologies of these debilitating neurological disorders.

Chapter 9 - Commonly known as starch, the $\alpha1$, 4 $\alpha1$, 6 gluco polisaccharide depends on its amylose: amylopectin ratio to understand their physic-chemical properties. Amylose is considered a long linear polymer with D-glucosyl units linked through by α D-1, 4 glucose linkages and amylopectin has a branched structure with poly-glucose residues linked by α D-1, 4 and α D-1, 6 glycose linkages. As an important prerequisite to evaluate the cassava starch granule, a carefully isolation of the starch in absence of a non degradative process is considered in this chapter in laboratory scale. The performance of the optimum starch molecular fractionation method gives an opportunity to obtain a best knowledge of its components. To determine the efficiency of fractionation starch two methods are presented, one method based on the differential solubility of starch components in a water- butanol saturated solution and another method based on complexing the starch molecule with Concanavalin A by its high affinity with carbohydrates. The composition and physical parameters give rise to diverse processing properties and therefore many uses of starch in non-food industries. In this chapter the cassava starch modification is described: 1) by gamma radiation to improve the water capacity property of a plastic biodegradable material, and 2) by esterification with fatty acid chloride to obtain a non polar agent with emulsifying capacity.

In: Cassava: Farming, Uses, and Economic Impact   ISBN:978-1-61209-655-1
Editor: Colleen M. Pace   © 2012 Nova Science Publishers, Inc.

Chapter 1

BIO-ETHANOL, BIO-PLASTICS AND OTHER FERMENTED INDUSTRIAL PRODUCTS FROM CASSAVA STARCH AND FLOUR

Ramesh C. Ray[1] and Manas R. Swain[2]*

[1]Central Tuber Crops Research Institute (Regional Centre),
Bhubaneswar 751 019, Orissa, INDIA
[2]Department of Biotechnology, College of Engineering
and Technology, Bhubaneswar-751003, Orissa, INDIA

ABSTRACT

Cassava is one of the richest source of starch. The tuberous roots contain up to 35% of starch and are low in proteins, soluble carbohydrates and fats that make starch extraction from cassava comparatively easier. Cassava starch and flour serve as raw materials for production of a number of industrial fermented products such as bio-ethanol, bio-plastics, acetone-butanol, dextrin, sugar (glucose and fructose) syrups and organic (lactic and glutamic) acids, microbial polysaccharides, baker's yeasts, etc. Bio-ethanol is produced from cassava starch and flour by liquefaction and saccharification, and subsequent fermentation by microorganisms. Cassava starch is converted

* Corresponding author: rc_ray@ rediffmail.com.

to lactic acid by lactic acid bacteria and polymerization of lactic acid by thermochemical reactions lead to the production of polylactic acid, which is then blended with other polymers to yield biodegradable plastics. In this chapter, the recent developments in the technology in the production of the above bio-products from cassava starch and flour have been discussed.

1. INTRODUCTION

Cassava (*Manihot esculenta* Crantz., Family: Euphorbiaceae) also known as tapioca or manioc is one of the major root crops grown in more than 100 countries of the humid tropics and sub-tropics. Globally cassava is grown in an area of 18.51 million ha producing 202.65 million tons with a productivity of 10.95 t/ha (FAO, 2009). It is one of the richest sources of starch. The roots (Figure 1) contain up to 35% starch (on fresh weight basis), and are low in proteins, soluble carbohydrates and fats that make starch extraction from cassava comparatively easier.

Figure 1. Cassava roots.

Unlike other tropical roots crops such as sweet potato, yams and aroids, 75% of cassava in Africa, 75-80% in Asia and 65% in Latin America are processed either as fermented foods and feeds or as industrial fermented products such as starch, sour starch, starch-based sweeteners (glucose, fructose and maltose syrups), ethanol, acetone, butanol, lactic and other organic acids, mono-sodium glutamate, microbial polysaccharides (xanthan, pullulan and scleroglucan), etc.

In this chapter, the recent developments in the technology in the production of industrial fermented products from cassava starch and flour have been discussed.

2. CASSAVA STARCH AND FLOUR

2.1. Cassava Starch

The roots of cassava should be processed within 24 h after harvest. The most essential factor in the production of good grade cassava starch is that the whole process, from harvesting the roots to completion of the final drying, should be carried out in the shortest time possible, since deterioration sets in from the time of root extraction from soil and proceeds throughout the process.

Basically, cassava starch manufacturing can be divided in to the following stages:

- Washing and peeling of the roots to remove and separate all adhering soil and as much protective epidermis as necessary.
- Rasping or disintegration to destroy the cellular structure and to rupture the cell walls to release the starch as discrete, undamaged granules from other insoluble matter.
- Screening or extraction to separate comminuted pulp into two fractions, viz. waste fibrous material (bagasse) and starch milk.
- Purification or dewatering to separate the solid starch granules from their suspension in water by sedimentation or centrifuging.
- Drying to remove sufficient moisture from the damp starch cake obtained during the separation stage so as to reduce the moisture content from 14 - 35% to 12 - 14%, a level low enough for long-term storage.
- Finishing operation such as pulverizing, sifting, and bagging.

The manufacture of cassava starch is carried out in mainly three types of establishments. The first is the cottage industry, where the work is carried out entirely by rudimentary hand tools, usually operated by a single family and producing 50 to 60 kg of crude starch per person per day. The second type, small scale enterprise, produces about 5 to 40 tons of roots per day, mainly because of more efficient rasping obtained by use of a prime mover of about 20 hp (horse power) and needing little skilled labor. The third type is the large scale factory which may sometimes operates from its own extensive plantations, thus assuring a regular supply of raw materials, processed using modern equipment. The third type of mill processes about 100 tons of cassava roots or more per day (Srinivas and Anantharaman, 2005).

2.2. Cassava Flour

As fresh cassava roots deteriorate within a few days after harvest, they are sliced in to small pieces or chips and dried in the sun or in hot-air oven. The dried chips can be preserved for months and processed in to cassava flour or meal. It is usually regarded as a difficult product to preserve in storage. Cassava flour is more hygroscopic than chips and starch. Probably fresh re-absorption of atmospheric moisture by the tiny particles is responsible for this. However, sometimes flour is preferred to storage of chips because besides a relatively higher bulk density and consequently lesser demand for space, some insect species, especially borers, which attack dried chips do not thrive in flour. Cassava flour rapidly adjusts to the humidity of the surrounding atmosphere. It has been reported that flour stored at 6.7% moisture content took up an additional 5% in 6 months when packed in cloth or gunny bags, but only 1 to 2% in polythene-lined sacks (Srinivas and Anantharaman, 2005).

3. INDUSTRIAL FERMENTATION PROCESSES

The industrial fermentation process can be broadly categorized into submerged (with soaking in water) and solid-state or solid-substrate (without soaking).

3.1. Submerged Fermentation (SmF)

SmF is the process in which the growth and anaerobic/partially anaerobic decomposition of the carbohydrates by micro-organisms in liquid medium occur with plenty availability of free water (Ray and Ward, 2006). Fermented industrial products like acetone-butanol, ethanol, lactic acid, etc. are the products of SmF. SmF is the process of choice for industrial operation due to the very well-known engineering aspects, such as fermentation modeling, bioreactor design and process control (Gutierrez- Correa and Villena, 2003).

In bioconversion of agricultural and food wastes to valuable chemicals, separation and purification of the products from the bulk liquid represent the highest percentages of the manufacturing cost. Therefore, the economic feasibility of reusing waste will strongly depend on the down-steaming processing efficiency. In recent years, considerable advances have been achieved in bimolecular purification technologies; one of the key goals is to achieve more selective, more efficient, and safer separation routes (Angenent et al., 2004). For example, a single- step direct lactic acid separation from fermentation broths has been successfully implemented using anionic fluidized- bed columns (Sosa, 2001). Superficial fluid technology (i.e. the use of fluid at a temperature and pressure that are greater than its critical levels) has also proven to be a useful tool from achieving separation of compounds directly from cultures, and is an attractive option because separation is usually performed at ambient temperatures using non-toxic, non-flammable solvents while achieving product crystallization (Angenent et al., 2004).

3.2. Solid-State Fermentation (SSF)

In contrast to SmF, solid-state fermentation (SSF) refers to the process where microbial growth and product formation occurs on the surface of solid materials. This process occurs in the absence of 'free' water, where the moisture is absorbed to the solid matrix (Pandey et al., 2000). Solid state fermentation has a series of advantages over submerged fermentation including lower cost, improved product characteristics, higher product yield, easiest product recovery and reduced energy requirement (Ray et al., 2008). Two main types of processes are used for the solid support in SSF. (1) SSF processes that use natural solid substrate like starch, flour or (ligno) cellulose residue or agro –industrial sources such as cassava, potato, sweet potato and pulp/ residues, etc.

In these cases substrates are also used as the source of carbon and nutrients for the microbial growth (Tengerdy and Szakaca, 2003). (2) SSF processes that use inert natural (cassava bagasse, sugarcane bagasse etc.) or artificial (perlite, amberlite, polyurethane foam and others) as solid supports. In these latter processes, support is used only as an attachment structure for the microorganisms (Ray et al., 2008).

4. CASSAVA FERMENTED INDUSTRIAL PRODUCTS

4.1. Bio- Ethanol

The concept of production of bio-ethanol from plant biomass and its use as energy sources is an age old one. Photo-biological energy conversion is the technique where carbon-based fuel can be obtained from plant sources and in the frantic search for alternative renewable resources certain sugar crops (sugar cane and sugar beet) and starch crops (cassava and maize) could also be identified as suitable for this use (Ward and Singh, 2005). Although the income elasticity of cassava is considered to be low, and although in term of ethanol production sugar crops do enjoy a better competitive position presently, in the future cassava can also become a strong contender for processing of ethanol (Toyin, 2000).

4.1.1. Cassava as a Bio-Ethanol Crop

The ability of cassava to compete with sugar crops for ethanol production will largely depend on the total production cost and this production cost can be cut considerably through increased use of cheaper raw material. Studies conducted earlier showed that for ethanol production from sugarcane or cassava, approximately 35% of the final cost is accountable to production cost and remaining 65% is the cost of raw material alone. As cassava is a drought resistant hardy crop and has the ability to grow well ever in poor and marginal lands that are generally considered unsuitable for other crops like sugarcane, it may in the long run enjoy comparable position, if not better as raw material source for an alcohol fuel economy (Debnath et al., 1990; Sherman, 2000). Cassava with the possibility of its storage as chip and flour extends the crop of operating the distilleries for a much greater length of the year, versus the sugarcane-based distilleries which are mostly seasonal in operation. Both cassava and sugarcane can be combined advantageously in alcohol fuel and many sugar mills can be utilized for an integrated fuel economy (Toyin, 2000). Like most biofuel crops, cassava has the potential to reduce carbon emissions. In addition, among the plants, cassava has the following special characteristics:

- It is an efficient converter of solar energy (250×10^3 k cal/ha), as it requires low inputs and yet, a high carbohydrate producer (Ray et al., 2010).

- As a drought-tolerant crop with multiple uses.
- It has a concentration of starch which varies between 16-35 %.
- It can be cultivated in sub-tropical and tropical climates.
- All components of the plant have economic value - the roots from cassava can be used as food, feed or bio-ethanol (from starch or flour), the leaves for forage, the fiber (cellulose) either as mulch or animal feed and with second generation technologies even for fuel (Ray *et al.*, 2008).
- Its bagasse, after starch extraction, has a higher biological value than the bagasse from sugarcane, when used as feed for animals or used for production of bio-ethanol.
- Its growing period (8-10 months) is similar to sugarcane (10-12 months), and the quantity of water required is one- third of sugarcane.
- It has some tolerance to salinity.

Therefore, based on the above characteristics, it seems that cassava is the most suitable plant for bio- ethanol production than other crops under hot and dry climatic conditions. In addition, possible use of bagasse as by-product of cassava include: burning to provide heat energy, paper or fiber board manufacturing, silage for animal feed or fiber for ethanol production (Ray *et al.*, 2008).

4.1.2. Bio-Processing for Ethanol Production

Production of ethanol from starch is not new in fermentation technology. Fresh cassava roots, starch or flour can be used for ethanol production. Chemically starch is a polymer of glucose (Ray *et al.*, 2010). Yeast cannot use starch directly for ethanol production. Therefore, cassava starch has to be completely broken down to glucose by a combination of two enzymes, viz., α-amylase and amyloglucosidase, before it is fermented by yeast to produce ethanol. The basic processes in the production of ethanol from cassava are somewhat cumbersome as compared to molasses. The step-wise processing of ethanol from cassava is given below (Ray *et al.*, 2008) (Figure 2):

1. Milling cassava chips or flour through sieve of 0.4 mm.
2. Cooking Process with enzymatic liquefaction and saccharification that include

a) Gelatinization
b) Liquefaction and dextrinization (using dilute inorganic acid or thermostable α-amylase)
c) Saccharification (using amyloglucosidase).

3. Fermentation, and
4. Distillation and dehydration

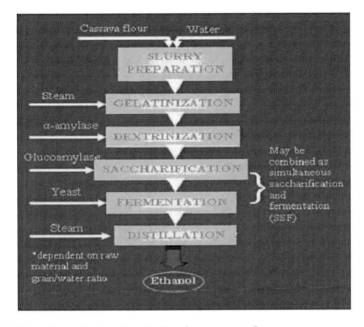

Figure 2. Flow-chart of ethanol production from cassava flour.

4.1.2.1. Milling

Crushing and milling of cassava chips are carried out by the following methods: (1) Dry milling - grinding of big solid particles into smaller ones (0.4 – 0.6mm), and (2) Wet milling - soaking in water. The dry milling has several advantages such as –

- Less capital investment in plant and equipment,
- Fewer control loops and simpler processing,
- Shorter time from construction to operation, and
- Minimal loss of starch.

4.1.2.2. Gelatinization

The crushed and milled flours are made into mash by steam cooking above starch gelatinization temperature (68-74^{0}C). The process is marked by melting of starch crystals, loss of birefringence and starch solubilization. Granules absorb large amount of water, swell to many times their original size, and open up enough for α-amylase to hydrolyze long chains into shorter dextrin (Ward *et al.*, 2006).

4.1.2.3. Liquefaction and Saccharification

The first step is generalization and conversion of starch to simpler sugar by a process called saccharification which is accomplished with the help of saccharifying agents like mild acid, amylase enzymes and substances containing amylase enzyme, e.g. malt. The hydrolysis process and the utilization of an efficient low-cost saccharifying agent are important factors in the production of ethanol from starch (Figure 3).

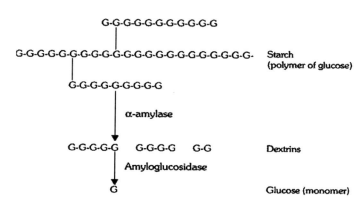

Figure 3. Enzymatic hydrolysis of cassava starch to glucose.

The starch is cooked first to release the starch granules, which are bound to the lingocellulosic compound of roots. These will also facilitate the reaction between the saccharifying agents (acids, enzyme, etc.) and the lower substrate. Cassava starch, having a lower swelling and generalization temperature can be easily saccharified to simple sugars with the help of acids or enzymes. The main advantages of cassava over any other energy crops are the presence of high fermentable sugar after saccharification. The use of dilute acid solution helps to recover approximately 98.8% of the reducing sugars from the stanch. The formation of a secondary reversion reaction which sets a limit to the yield of glucose that may be obtained, the formation of high inorganic salts due to pH adjustment and the corrosion of the machinery are some of the disadvantages of the use of acids in the saccharification process.

The enzyme amyloglucosidase (AMG) (glucoamylase) derived commercially from the strains of molds *Aspergillus* and *Rhizopus* can hydrolyze gelatinized starch completely to glucose units. It acts by cleaving the α-1-4 glycosidic linkage of the non-reducing terminal glucose unit. It can also hydrolyze the α-1-6 linkage from the amylopectin. Malt contains the three most important enzymes for the starch breakdown, namely, α- amylase, β-amylase, and amyloglucosidase. The α- amylse splits the α- 1-4- link randomly within the molecules, forming dextrins, which are small chain of glucose. This makes the generalized starch slurry more fluid and supplies more chain ends for the action of saccharifying enzymes. Since all the α- amylase can not split the α-1-6- links of starch and dextrins, all the branch points remain intact after α-amylase action. The amylase also breaks the α-1-4 link of dextrin and starch, but only from the non-reducing ends of the molecules, resulting to maltose formation. Since neither of these enzymes attack α-1-6 linkage, their combined action converts only up to 85% of the starch in to reducing sugars. Amyloglucosidase splits the α-1,6 linkage of the maltose and short-chain of carbohydrates, thus completing the hydrolysis of starch into fermentable sugars. The continued action of the above enzymes produced from the microorganisms can also split the cassava starch to glucose units.

Since amylase enzymes do not tolerate temperature above 55°C, they must be added after gelatinization has taken place. The use of heat-stable bacterial amylase has received considerable attention at least for effective conversion of starch to sugar. Commercial enzymes i.e. thermostable α-amylase (Termamyl®) and amyloglucosidase are consistently used to liquefy and saccharify the cassava starch to fermentable sugars (Zanin and de Morase, 1998; Paolucci *et al.*, 2000), achieving an ethanol yield of 400-450 litre/ton of the cassava starch, thus processing a total of 3,969 million litres of ethanol per annum which is 88% efficiency of the installed capacity (Toyin, 2000). Newer enzymes such as Spezyme® and Stargen ™ were found more effective than Tarmamyl® and AMG in saccharifying cassava starch for production of ethanol (Shanavas *et al.*, 2011).

4.1.2.4. Fermentation

4.1.2.4.1. Microorganisms

Traditionally, the yeast *Saccharomyces cerevisiae* is used for ethanol production. In recent years, however, research is being focused in processes involving on Gram negative anaerobic bacterium, *Zymomonas mobilis* as it has several advantages (Nellaiah and Gunasekaran, 1992, Panesar *et al.*, 2001). Some improvements over the traditional process of ethanol production are only cited here. Ethanol production from fresh cassava roots, flour or starch using *S. cerevisiae* showed 90-95% fermentation efficiency when the slurry consisting 20% solids was hydrolyzed by liquefaction- saccharification processes. Saccharification of the mash resulted in 10-12% reducing sugar and inoculation at this stage with yeast led to simultaneous saccharification and fermentation and hence curtails the total time of the fermentation process (Ray *et al.*, 2004).

Direct ethanol production from raw sago starch was investigated using a mixture of strains of *Aspergillus niger* (Pranamuda *et al.*, 1994) and ethanol yeast, *S. cerevisiae*. Ethanol yield from raw sago starch, using the fed-batch mixture, was 50 ml ethanol/100g starch (Pranamuda *et al.*, 1994).

4.1.2.5. Fermentation Conditions

Large volumes of the saccharified starch are fed into fermentation vessels and inoculated with actively growing *S. cerevisiae*. Many strains of yeasts with varying capabilities and tolerances have been reported. The yeast inoculation is usually 5 to 10% of the total volume of the fermentation medium. The optimum concentration of sugars for ethanol fermentation is 12 to 18 %. The pH of the mash for fermentation is optimally 4 to 4.5 and the temperature of fermentation is 28 to 32°C. Sugar is converted to ethanol, carbon dioxide and yeast/bacterial biomass as well as much smaller quantities of minor end products such as glycerol, fusel oils, aldehydes and ketones (Jacques *et al.*, 1999; Laopaiboon *et al.*, 2007).

4.1.2.6. Distillation and Dehydration

Alcohol is recovered from the fermented mash after 48 to 72 h at the end of fermentation; the yeast is separated from the mash by centrifugation or sedimentation and used for the next batch of fermentation. The resulting liquid is distilled for the recovery of ethanol.

Alcohol distilled from fermented mash is concentrated up to 95% v/v. This is further concentrated to produce ethanol with 99.6% v/v (minimum) concentration. The treatment of vinasse generated in the distillation section can be done using following option: concentration of part of vinasse to 20 to 25% solids followed by composting using press mud available and concentration of rest of the vinasse to 55% solids and can be used as liquid fertilizer.

4.1.2.7. Problem Associated with Fermentation

The conventional batch fermentation requires high cost manpower, because permanent supervision is required and the preparation procedure is elaborate and has to be repeated within small interval of time. The shorter time of fermentation reduces the danger of contamination, but requires an uneconomic interruption of the process. The low production of ethanol is one of the serious disadvantages of batch fermentation. The yields of batch fermentation which is mostly performed in alcoholic destillaries are about 6% by volume for fermentation time of 36h. The productivity may be explained by the fact that physiological properties of the microorganisms are not sufficiently considered in batch fermentation. The result is a low propagation rate of the cells. High substrate concentration at the beginning and the increasing product concentration at the end of the fermentation process cause inhibition of growth.

4.1.2.8. New Trends in Cassava Bio-Ethanol Technology

The problem of substrate inhibition and product inhibition could be solved by adopting new trends in bio-ethanol technology

4.1.2.8.1. Continuous Fermentation

This is characterized by continuous addition of nutritive media and by continuous removal of cells and fermenter solution. The initial growth of the cell is identical with that in batch fermentation. When a favorable exponential growth phase is obtained, the organisms may be kept permanently in this stationary phase by regulation of influx and efflux. The whole system is thus in a steady state. The yeasts are kept for a very long time in an exponential growth phase; a higher substrate transformation rate may be obtained in a very short time. As a result of this increased productivity, even small plants are economical. The danger of contamination and the possibility of mutation of the organisms are some of the problems encountered in continuous fermentation (Ray and Edison, 2005).

4.1.2.8.2. Vacuum Fermentation

Vacuum Fermentation is done is a simple stirred tank fermenter. In producing a vacuum (32 to 35 mm/kg) in bioreactor, a mixture of ethanol and water is distilled directly at 30°C from the fermenter. The yeast cells are not hampered by this procedure; in contrast, the product inhibition occurring at ethanol concentration of maximum 6% in the medium is overcome. The technical application of vacuum fermentation, however, has become questionable because of the high energy requirements of the cooking and pumping system (Ray and Edison, 2005).

4.1.2.8.3. Immobilized Yeast Cells

The most recent advances in alcohol production technology are the application of immobilized cells in continuous column reactor. The approach combines several advantages:

- High cell concentration can be obtained;
- The reaction rate is accelerated;
- Operation can be performed at a high dilution rate without a wash out,
- The end product inhibition is eliminated, since alcohol is removed continuously;
- Anaerobic condition can be achieved. Since cells are entrapped into gel matrix; and
- Costly equipment designs such as fermenters, agitator will be eliminated.

In recent years, immobilization of enzymes or microbial cells or co-immobilization of both for the simultaneous starch saccharification and fermentation (SSSF) process has attained scientific and technical interest for alcoholic fermentation. By concentrating an active cell biomass or enzyme in a bioreactor, the efficiency of bioconversion increases, as does the reactor productivity which, in turn, results in the reduction of the reactor size for a given production rate (Nunez and Lema, 1987). Immobilization can be carried out in different ways; adsorption and entrapment in matrices are the methods most commonly used (Kim and Rhee, 1993; Yu et al., 1996). Among the various immobilized methods tested, a co-immobilized system using Z. mobilis in chitin or sodium alginate appeared most promising with respect to ethanol productivity from cassava sago starch (Kim and Rhee, 1993; Nellaiah and Gunasekaran, 1992). However, despite many efforts undertaken worldwide, Zymomonas has not yet been commercialized due to its low tolerance to temperature, ethanol and limited substrate range.

4.1.2.8.4. Simultaneous Starch Saccharification and Fermentation (SSSF)

The most important process development made for enzymatic hydrolysis of various biopolymers (i.e., starch, cellulose and hemi-cellulose) - containing crops and biomass is the introduction of simultaneous starch saccharification and fermentation (SSSF) process (Ray and Naskar, 2008). However, in SSSF both saccharifying enzyme and fermenting microorganisms are applied simultaneously. As the conversion of starch, cellulose and hemi-cellulose into sugars is processed by respective microbes or their enzymes, the fermentative organisms convert them into ethanol.

The SSSF process combines enzymatic hydrolysis of starch or flour to glucose with ethanol fermentation in a single operation using the organism, *Z. mobilis*. Glucose is fermented to ethanol immediately after saccharification rather than accumulated in this system. As a consequence, this process offers a great potential of increased rate of hydrolysis, reduced inhibitory effect of substrate and decreased capital cost (Amutha and Gunasekaran, 2000). A Pilot scale study was successfully conducted for ethanol fermentation by *Z. mobilis* using SSSF process in Korea (Kim *et al.*, 1990; 1992). One of the problems associated with this process, however, is that the optimum temperature for the enzyme, AMG is appreciably higher (65°C) than that for fermentations (30-35°C). The lower temperature used for the SSSF process, therefore, requires increased enzyme levels to maintain similar productivity.

4.1.2.9. Most Recent Studies

Cassava pulp (bagasse) was hydrolyzed with acids or enzymes. A high glucose concentration (>100 g/L) was obtained from the hydrolysis with 1 N HCl at 121 °C, 15 min or with cellulase and amylases. While a high glucose yield (>0.85 g/g dry pulp) was obtained from the hydrolysis with HCl, enzymatic hydrolysis yielded only 0.4 g glucose/g dry pulp. These hydrolysates were used as the carbon source in fermentation by *Rhizopus oryzae* NRRL395. *R. oryzae* could not grow in media containing the hydrolysates treated with 1.5 N H_2SO_4 or 2 N H_3PO_4, but no significant growth inhibition was found with the hydrolysates from HCl (1 N) and enzyme treatments. Higher ethanol yield and productivity were observed from fermentation with the hydrolysates when compared with those from fermentation with glucose in which lactic acid was the main product. This was because the extra organic nitrogen in the hydrolysates promoted cell growth and ethanol production (Thongchul *et al.*, 2010).

Raw cassava starch was transformed into ethanol in a one-step process of fermentation, in which are combined the conventional processes of liquefaction, saccharification, and fermentation to alcohol. *Aspergillus awamori* NRRL 3112 and *Aspergillus niger* were cultivated on wheat bran and used as Koji enzymes. Commercial *A. niger* amyloglucosidase was also used in this experiment. A raw cassava root homogenate-enzymes-yeast mixture fermented optimally at pH 3.5 and 30°C, for 5 days and produced ethanol. Alcohol yields from raw cassava roots were between 82.3 and 99.6%. Fungal Koji enzymes effectively decreased the viscosity of cassava tuber fermentation during incubation. Commercial *A. niger* amyloglucosidase decreased the viscosity slightly. Reduction of viscosity of fermentation mashes was 40, 84, and 93% by commercial amyloglucosidase, *A. awamori*, and *A. niger* enzymes, respectively. The reduction of viscosity of fermentation mashes is probably due to the hydrolysis of pentosans by Koji enzymes (Ueda *et al.*, 1981).

4.1.2.10. Cassava Bagasse for Ethanol Production

Cassava bagasse constitutes about 15-20% by weight of the processed cassava chips/roots, is retained on sieves during the rasping process. It contains about 55-65% starch (on dry weight basis) (Sriroth *et al.*, 2000; Jyothi *et al.*, 2005). Cassava bagasse was used as solid substrate (support and nutrient source) for production of various fermented products such as enzymes, ethanol, lactic acid, etc (Ray *et al.*, 2008).

A novel technique, involving the hydrolysis of starch present in cassava bagasse in shallow layers in stainless steel trays, is developed to facilitate the use of higher slurry concentrations. The use of slurry containing 30% solids, 4% sulphuric acid, 30 min saccharification time at 121°C resulted in the complete conversion of starch into reducing sugars. The spent residue, after separation of the hydrolysate, contained about 24% of the total sugars formed and these were recovered to the extent of 90-94% by using a counter-current extraction technique. A large scale saccarification of a 75 kg batch gave 75 litre pooled hydrolysate containing 15% reducing sugars. An overall process efficiency of 76.4% was observed with the fermentation of hydrolysate pooled with the counter-current extract for alcohol production. However, fermentation of the whole saccharified pulp without the separation of the hydrolysate and acid-enzyme hydrolysis of the waste gave lower efficiencies (Srikanta *et al.*, 1987).

In an earlier study, Fermentation studies for the production of ethanol on saccharified cassava bagasse revealed that enrichment with mineral salts and nitrogen was essential. Inoculum size of 24.2 x 10^6 cells/ 50 ml medium was optimal while highest quantity of alcohol was formed at 96 h of fermentation. Though optimum initial sugar concentration in fermentation medium was found to be about 15%, the saccharified waste needed expensive concentration treatment to raise the sugar concentration to optimum level. Economic considerations suggest the use of saccharified waste without concentration

(Kunhi *et al.*, 1981). In a subsequent study, a total of 16.5% reducing sugars in the saccharified pulp of cassava bagasse were achieved with the use of 30% slurry. The yield of ethanol was highest and the amount of residual reducing sugars was lowest with the use of 2.5% acid. The increase in dose of glucoamylase leads to improved yields of ethanol without any lowering in the residual reducing sugars. The ethanol yield and productivity were better and the residual reducing sugars were lower in solid phase fermentation as compared to the fermentation of liquid hydrolysate obtained by hydraulic pressing of the saccharified pulp. The slightly lower yield of ethanol in large batch static fermentation probably due to poor mass transfer and limited contact of yeast cells as well as enzyme with their substrates could be effectively overcome by employing appropriate strategies (Jaleel *et al.*, 1988)

The peak ethanol yield of 232 g /kg cassava bagasse shows that nearly 80% conversion of fermentable sugar into ethanol has been achieved assuming 90% conversion of starch in cassava bagasse to fermentable sugars (Ray *et al.*, 2008). Further, highest ethanol yield was observed at 60% moisture holding capacity and 120h of fermentation

In a most recent study, an alternative cassava bagasse saccharification process, which utilized the multi-activity enzyme from *Aspergillus niger* BCC17849 and obviated the need for a pre-gelatinization step, was developed. The crude multi-enzyme composed of non-starch polysaccharide hydrolyzing enzyme activities, including cellulase, pectinase and hemicellulase acted cooperatively to release the trapped starch granules from the fibrous cell wall structure for subsequent saccharification by raw starch degrading activity.

A high yield of fermentable sugars, equivalent to 716 mg glucose and 67 mg xylose/g of cassava bagasse, was obtained after 48 h incubation at 40 °C and pH 5 using the multi-enzyme, which was greater than the yield obtained from the optimized combinations of the corresponding commercial enzymes. The multi-enzyme saccharification reaction can be performed simultaneously with the ethanol fermentation process using a thermo-tolerant yeast, *Candida tropicalis* BCC7755.

The combined process produced 14.3 g/L ethanol from 4% (w/v) cassava bagasse after 30 h of fermentation. The productivity rate of 0.48 g/L/h was equivalent to 93.7% of the theoretical yield based on total starch and cellulose, or 85.4% based on total fermentable sugars. The non-thermal enzymatic saccharification process described was found more energy efficient and yielded more fermentable sugar than the conventional enzymatic process (Rattanachomsri *et al.*, 2009).

4.2. Acetone-Butanol

The acetone butanol fermentation is an anaerobic fermentation brought about by *Clostridium acetobutylicum* and closely related species or variants. The important products produced by these organisms are butyl alcohol, acetone and ethyl alcohol. The primary carbohydrate raw materials for the butanol-acetone fermentation processes are corn, corn black strap molasses and high test molasses. A number of other raw materials indicating cassava starch have been tried and successfully used for the acetone butanol fermentation. Gelatinized starch with the addition of nitrogenous and phosphatic nutrients can serve as an excellent medium for the acetone-butanol fermentation. Ammonia may be supplied in the form ammonium sulfate or as aqueous ammonium hydroxide. Generally phosphate is applied as superphosphate. The temperature should be maintained at 35° to 37°C and after a fermentation period of 16h with *Clostridium acetobutylicum*, significant increase in the ratio of butanol to other solvents resulted (Maddox *et al.*, 1995).

4.3. Lactic Acid and Bio- Plastic

There are four major categories for the current uses and applications of lactic acid: food, cosmetic, pharmaceutical, and chemical applications. The potential applications of lactic acid are illustrated in Figure 4. Since lactic acid is classified as GRAS (Generally Regarded As Safe) for use as a food additive by the US FDA (Datta *et al.*, 1995), it is widely used in almost every segment of the food industry, where it serves in a wide range of functions, such as flavoring, pH regulation, improved microbial quality, and mineral fortification. Moreover, lactic acid is used commercially in the processed meat and poultry industries, to provide products with an increased shelf-life, enhanced flavor, and better control of food-born pathogens. Due to the mild acidic taste of lactic acid, it is also used as an acidulant in salads and dressings, baked goods, pickled vegetables, and beverages. Currently, lactic acid is considered the most potential feedstock monomer for chemical conversions, because it contains two reactive functional groups, a carboxylic group and a hydroxyl group. Lactic acid can undergo a variety of chemical conversions into potentially useful chemicals, such as propylene oxide (*via* hydrogenation), acetaldehyde (*via* decarboxylation), acrylic acid (*via* dehydration), propanoic acid (*via* reduction), 2,3-pentanedione (*via* condensation), and dilactide (*via* self-esterification) (Bhatia *et al.*, 2007).

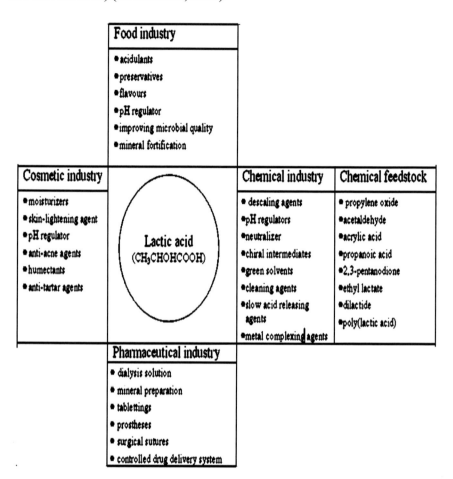

Figure 4. Diagram of the commercial uses and applications of lactic acid and its salt (Wee et al., 2006).

Lactic acid serves as a feedstock monomer for the production of poly lactic acid (PLA), which serves as a biodegradable commodity plastic. PLA, linear aliphatic polyester, produced from renewable resources has attracted much attention in recent years due to its biodegradability and could be a possible solution for solid waste (Ray and Bousmina, 2005). PLA has good mechanical, thermal and biodegradable properties and therefore is a good polymer for various end-use applications (Qi and Hanna, 1999). However, other properties like flexural properties, heat distortion temperature (HDT), gas permeability, impact strength, melt viscosity for processing, etc. are not good enough in applications like packaging (Ogata *et al.*, 1997). Also high price and brittleness of PLA lowers the possibility of its commercialization. Therefore, blending PLA with other suitable biodegradable polymer such as cassava starch which has comparably better flexural properties, excellent impact strength, melt possibility will modify various properties and also contribute towards low overall material cost. PLA was reported to be miscible with other stereo isomers such as poly (DL-lactic acid) (Tsuji and Ikada, 1996), and some other polymers like poly (ethylene oxide) (PEO) (Nijienhuis *et al.*, 1996), poly (vinyl acetate) (PVA) (Gajria *et al.*, 1996), and poly (ethylene glycol) (PEG) (Sheth *et al.*, 1997). The blends had varying properties of the blended polymers according to their ratio of mixing. The optically pure lactic acid can be polymerized into a high molecular mass PLA through the serial reactions of polycondensation, depolymerization, and ring-opening polymerization (Södergård and Stolt, 2002). The resultant polymer, PLA, has numerous uses in a wide range of applications, such as protective clothing, food packaging, mulch film, trash bags, rigid containers, shrink wrap, and short shelf-life trays (Drumright *et al.*, 2000, Vink *et al.*, 2003). The recent huge growth of the PLA market will stimulate future demands on lactic acid considerably (Datta *et al.*, 1995, Lunt, 1998).

Microorganisms that can produce lactic acid can be divided into two groups: bacteria and fungi (Litchfield, 1996). The microorganisms selected for recent investigations of the biotechnological production of lactic acid are listed in Table 1.

In the following paragraph, the research done on lactic acid production from cassava starch, flour and bagasse are discussed. L-lactic acid production by *Rhizopus oryzae* from raw cassava starch as a sole carbon source was studied. The maximum L-lactic acid production in shake flasks was 68.32 g/L on day 5 with the shaking speed of 200 rpm at 30°C. L – lactic acid yield and productivity were 0.59 g/g substrate and 0.57 g/L/h, respectively. In a jar fermentor, the maximum L-lactic acid production was 54.62 g/L on day 4 of cultivation at the agitation speed of 400 rpm and the aeration rate of 1.5 vvm. The L – lactic acid yield was 0.49 g/g substrate with the productivity of 0.56 g/L/h (Naranong and Poocharoen, 2001).

L-lactic acid was produced from raw cassava starch, by simultaneous enzyme production, starch saccharification and fermentation in a circulating loop bioreactor with *Aspergillus awamori* and *Lactococcus lactis* spp. *lactis* immobilized in loofa sponge. *A. awamori* was immobilized directly in cylindrical loofa sponge while the *L. lactis* was immobilized in a loofa sponge alginate gel cube. In the loofa sponge alginate gel cube, the sponge serves as skeletal support for the gel with the cells. The alginate gel formed a hard outer layer covering the soft porous gel inside. By controlling the rate and frequency of broth circulation between the riser and downcomer columns, the riser could be maintained under aerobic condition while the downcomer was under anaerobic condition. Repeated fed-batch L-lactic acid production was performed for more than 400 h and the average lactic acid yield and productivity from raw cassava starch were 0.76 g lactic acid/ g starch and 1.6 g lactic acid/ L/ h, respectively (Roble *et al.*, 2003).

Table 1. Microorganisms used for production of L- lactic acid

Organism
Rhizopus oryzae ATCC 52311
Rhizopus oryzae NRRL 395
Enteroccus faecalis RKY1
Lactobacillus rhamnosus ATCC 10863
Lactobacillus helveticus ATCC 15009
Lactobacillus bulgaricus NRRL B-548
Lactobacillus casei NRRL B-441
Lactobacillus plantarum ATCC 21028
Lactobacillus pentosus ATCC 8041
Lactobacillus amylophilus GV6
Lactobacillus delbrueckii NCIMB 8130
Lactobacillus lactis ssp. *lactis* IFO 12007
Lactobacillus plantarum MTCC 1407

Immobilized *Lactobacillus delbrueckii* cells produced lactic acid up to six batches from cassava bagasse without any decline in lactate production. Lactate yield for *L. delbrueckii* was 0.93 g lactic acid / g reducing sugar with a production rate of 0.33 g/ L/h (John *et al.*, 2006).

A column bioreactor packed with immobilized *L. delbrueckii* was run for three weeks in a continuous mode with a lactate yield of 0.75-0.95 g lactic acid/g reducing sugar with an average production rate of 0.48 g/L/h (John *et al.*, 2007).

Ray *et al.*, (2009) used cassava bagasse in semi solid-state fermentation for the production lactic acid by using *Lactobacillus plantarum*. Response

Surface Methodology was used to evaluate the effect of main variables, i.e. incubation period, temperature and pH on lactic acid production. The experimental results showed that the optimum incubation period, temperature and pH were 120 h, 35° C and 6.5, respectively. Maximum starch conversion by *Lactobacillus plantarum* MTCC 1407 to lactic acid was 63.3%. The organism produced 29.86 g of (L+) lactic acid from 60 g of starch present in 100 g of cassava bagasse. The LA production yield (i.e. mass LA produced mass /starch present in cassava bagasse x 100) was 49.76%. Coelho *et al* (2010) studied effects of different medium components added in cassava wastewater for the production of L (+)-lactic acid by *Lactobacillus rhamnosus* B 103. The use of cassava wastewater (50 g /L of reducing sugar) with Tween 80 and corn steep liquor, at concentrations (v/v) of 1.27 ml /L and 65.4 ml /L, respectively led to a lactic acid concentration of 41.65 g/ L after 48 h of fermentation. The maximum lactic acid concentration produced in the reactor after 36 h of fermentation was 39.00 g/ L using the same medium, but the pH was controlled by addition of 10 ml /L NaOH (Coelho *et al.*, 2010).

To reduce the production cost of biodegradable plastics, the fermentation performance of L-lactic acid for a new fermentation medium, fresh cassava roots (FCRs) as a substrate slurried with tofu liquid waste (TLW) as basal medium, was investigated by batch fermentation of *Streptococcus bovis*. The fermentation properties of the three substrates, namely, FCR, cassava starch and glucose, which were independently mixed with TLW, were compared with those independently mixed with the standard basal medium, trypto-soya broth (TSB). Experiments were conducted at various sugar concentrations of the substrates with $CaCO_3$ as a neutralizer. The maximum L-lactic acid concentrations (C_{La}) obtained using the three substrates in TLW were about 75% of those obtained using TSB caused by less nutrients in the TLW. The L -lactic acid productivities (P_{La}) and the specific growth rates of *S. bovis* (μ) in TLW were about 1/4 to 1/3 and 1/5 to 1/4 of those in TSB, respectively. The maximum C_{La}, P_{La} and μ were obtained at 10% w/w sugar concentration. Total yields (η) were nearly constant up to 10% w/w sugar concentration for TSB and TLW, that was, 80% to 85% and 50% to 60%, respectively. But their total yields decreased in more than 10% w/w sugar concentration in both basal media, because of substrate inhibition. The fermentation properties (C_{La}, P_{La}, μ, and η) were found to be in the order of: FCR > cassava> glucose for all concentrations of the three substrates. The fermentation properties for FCR and cassava were higher than those for glucose, in TLW or TSB, because *S. bovis* in a medium containing starch (FCR and cassava) had more amylase activity than in a medium containing glucose. The nutrients in FCR with poor nutrient basal medium (TLW) more strongly affected the fermentation properties than those in FCR with rich nutrient basal medium (TSB) (Ghofar *et al.*, 2005).

Gelatinized cassava bagasse was enzymatically hydrolyzed and starch hydrolysate containing reducing sugar was used to moisten the inert sugarcane bagasse, which was used as the solid support for solid state fermentation. This substrate was supplemented with 0.5 g/5 g support NH_4Cl and yeast extract. SSF was carried out in 250 ml Erlenmeyer flasks at 37 °C using *Lactobacillus delbrueckii* as inoculum. Key parameters such as initial moisture content and initial sugar were optimized statistically by response surface methodology. A maximum of 249 mg/gds (gram dry substrate) L-lactic acid was obtained after 5 days of fermentation under the optimized conditions with a conversion efficiency of about 99% of the initial reducing sugars (John *et al.*, 2006).

4.4. Organic Acids

4.4.1. Citric Acid

With an estimated annual production of about 1,000,000 tons, citric acid is one of fermentation products with the highest level of production worldwide (Soccol *et al.*, 2003). Considerable amounts of citric acid are required in several industrial processes. The food industry consumes about 70% of the total citric acid production, while other industries consume the remaining 30% (Pandey *et al.*, 2000). Commercially, production of citric acid is generally by submerged fermentation of sucrose or molasses using the filamentous fungus *Aspergillus niger* (Vandenberghe *et al.*, 2000).

Citric acid production using cassava bagasse was studied in different bioreactors. The best results (26.9 g/100g of dry cassava bagasse) were obtained in horizontal drum bioreactor using 100% gelatinized bagasse, although the tray-type bioreactor offers advantages and showed promise for large-scale citric acid production in terms of processing costs (Prado *et al.*, 2005).

4.4.2. Itaconic Acid

The possibility of production of itaconic aid, an unsaturated dicarboxylic acid, from cassava starch was successfully demonstrated from *Aspergillus itaconicus*. Three days old spore suspension of *A. itaconicus* was inoculated in to cassava starch medium containing 4% cassava starch and 0.25% NaCl. The pH of the medium was adjusted to 4.5. Two ml of concentrated rice bran liquor made by boiling 500g rice bran soaked in 100 ml of water for 30 min was added to the medium. The spores of *A. itaconicus* harvested from the growth medium after 3 days of incubation were used for inoculation of the production medium. The concentration of the inoculums was in the ratio of 1:10. After 15 days of incubation, the fermented liquor was concentrated under a vacuum to one tenth of its original volume. The itaconic acid was crystallized from the hot concentrated liquor filtered and dried. Cassava starch could thus be utilized as a substitute for sugar in the production of itaconic acid as carbon source (Balagopalan *et al.*, 1988)

4.4.3. Monosodium Glutamate (MSG)

MSG is an important flavor enhancer of a wide range of savoury foods. China is the largest producer and consumer of MSG in the world. The starch has first to be degraded to sugars, which are then converted by microorganisms such as *Brevibacterium glutamicum* to glutamic acid. This is then converted to MSG salt (Jiang *et al.*, 1993).

4.5. Sugar Syrups

The conversion of starch into a range of syrups and other derivatives (Figure 5) is becoming increasingly important in some countries where these products can be used to replace more expensive imported sugar extracted from cane or beet. These conversions employ a sophisticated technology based on microbial enzymes and can utilize any starch source including sweet potato and cassava, which may be particularly appropriate as it is highly susceptible to saccharification by enzymes (Paolucci *et al.*, 2000). Though starch can be converted into sugars by the use of acids, this method is rapidly giving way to the use of immobilized bacterial enzymes (Nunez and Lema, 1987; Zanin and de Morase, 1998), the specific properties of which give rise to a variety of compounds useful in the dessert, bread, fermented milk products, brewing and other industries. Glucose syrup, for example, is produced from starches, including sweet potato and cassava starch by bacterial amylase. However, it has only 70% of the sweetness of sucrose.

High fructose syrup (HFS) is a highly valued liquid sweetener for beverage, confectionery and processed food industry, owing to its special attributes like high solubility and non-crystalline nature. Even though 85% HFS production is from corn, increased food demand has necessitated the search for alternative substrates and starchy root crops like cassava and sweet potato are potential raw materials. However, the economic production needs direct use of the roots and simplification of the cost-intensive steps. Glucose yield was compared from six treatment systems viz., liquezyme–dextrozyme (T1), Stargen (T2), Stargen in two split doses (T3), Spezyme–Stargen (T4), Stargen (60 °C;T5) and Spezyme–Stargen (60 °C; T6). Glucose was higher (22–25%) from cassava than sweet potato (14.0–15.7%), owing to the high starch content in cassava. Conversion to glucose was higher in T1–T4 (95–98%) compared to 88–92% for T5 and T6. Although the fructose yield was more from cassava (8.36–9.78%) than sweet potato (5.2–6.0%), percentage conversion was similar (37–38%) for both the roots. The cost of production of HFS could be reduced by the direct hydrolysis of root slurry using Stargen (Jhanson *et al.,*2009).

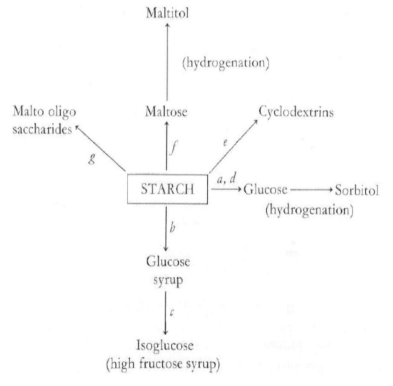

Figure 5. Conversion of starch into syrups and other derivatives from cassava roots. a, based on Ray and Ward (2006); b, α-amylase; c, glucose isomerase; d, glucoamylase; e, cyclodextrin glucanotransferase; f, β-amylase; g, maltooligosaccharide-forming amylases.

The partial conversion of glucose into its isomer fructose by bacterial glucose isomerase to give high fructose syrup, or iso-glucose, as it is also known, gave rise to a substance with much greater sweetness than sucrose. Iso-glucose (which contains at least 42% fructose) has become an important replacement for sucrose in several areas of the food industry where it can be used in lower concentration than sucrose to provide the same sweetness, hence reducing the energy (calorie) content of the food. A further sugar, is maltose produced from starch by the action of β- amylase (Fontana *et al.*, 2001) is useful to the brewing industry. Two novel products, which have been developed from cassava starch, are cyclodextrin and maltooligosaccharides (Raja *et al.*, 1990; Fontana *et al.*, 2001). Cyclodextrin has a variety of uses including the stabilization to heat, light, or oxygen of compounds such as fatty acids and vitamins. These products, which are expensive, could be produced from cassava and sweet potato starch thus adding to its value.

4.6. Microbial Polysaccharides

Cassava bagasse serve as substrate for production of microbial exo-polysaccharides, which have a number of uses in pharmaceutical, cosmetic, brewing and electronics industries (Roukas, 1998). Cassava bagasse (Ray and Moorthy, 2007) and cassava flour (Selbmann *et al.*, 2002) have been used to produce microbial polysaccharides (yeast like fungus *Aureobasidium pullulans*) such as pullulan and xanthan (by the bacterium, *Xanthomonas campestris*).

4.7. Baker's Yeast and Other Products

Baker's yeast (*Saccharomyces cerevisiae*) was cultured on hydrolyzed waste cassava starch (Ejiofor *et al.*, 1996). Volatile compounds are produced by the fungus *Geotrichum fragrans* using cassava waste water as the substrate (Damasceno *et al.*, 2003).

CONCLUSIONS AND FUTURE PERSPECTIVES

Starch and starch-derived products constituted as the major industrial produces form cassava. This chapter has focused exclusively on the fermented industrial products like ethanol, lactic and other organic acids, sugar syrups, amino acid derivatives, etc, which are developed by fermentation of either cassava starch or flour by various microorganisms. Cassava is being considered as the raw material for production of ethanol. An enormous quantity of ethanol is required as either substitute or 10% additive to gasoline in many parts of the world. Countries like Brazil, China, Thailand and The Philippines are the leaders in using alcohol-driven motor vehicles. Besides, ethanol is also required for the production of a variety of industrial chemicals, such as cellulose triacetate, vinyl acetate, poly vinyl chloride, styrene and polystyrene. Likewise, polylactic acid, another fermented product from cassava starch and flour has tremendous international market as raw material for biodegradable plastics. The very high starch content (< 20-30% on dry weight basis), high productivity (13-18 tons yield/ha in marginal and arable land and up to 60 tons yield/ha in irrigated land) and high tolerance of the crop to draught, salinity and vagaries of climatic changes have valued cassava as an ideal industrial crop for bio-processing and value-addition for industrial chemicals and other bio-products.

REFERENCES

Angenent, L.T., Karim, K., Al-Dahhan, M.H., Wrenn, B.A. and Dimiguez-Espinosa, R. (2004). Production of bioenergy and biochemicals from industrial and agricultural waste water. *Trends Biotechnol.* 22: 477- 485.

Amutha, R. and Gunasekaran, P. (2001). Production of ethanol from liquefied cassava starch using co-immobilized cells of *Zymomonas mobilis* and *Saccharomyces diastaticus. J. Biosci. Bioeng.* 92:560-564.

Balgopalan, C., Padmaja, G., Nanda, S.K., and Moorthy, S.N. (1988). *Cassava in Food Feed and Industry.* CRC Press. Boca Rato, Florida.USA, pp.205.

Bhatia1, A., Gupta1, R. K., Bhattacharya1 S. N. and Choi, H. J. (2007). Compatibility of biodegradable poly (lactic acid) (PLA) and poly (butylene succinate)(PBS) blends for packaging application. *Kor.-Aust. Rheol. J.* 19: 125-131.

Coelho, L. F., de Lima, C.J. B., Bernardo, M.P., Alvarez, G. M. and Contiero, J. (2010). Improvement of L(+)-lactic acid production from cassava wastewater by *Lactobacillus rhamnosus* B 103. *J. Sci. Food Agric.* 90: 1944-1950

Damasceno, S., Cereda, M.P., Pastone, G.M. and Oliveira, J.G. (2002). Production of volatile compounds by *Geotrichum fragrans* using cassava waste water as substrate. *Process Biochem.* 39 (3): 411-415.

Datta, R., Tsai, S.P., Bonsignore, P., Moon, S.H. and Frank, J.R. (1995). Technological and economic potential of poly (lactic acid) and lactic acid derivatives. *FEMS Microbiol. Rev.* 16: 221–231.

Debnath, S., Banerjee, M. and Majumdar, S.K. (1990). Production of alcohol from starch by immobilized cells of *Saccharomyces diasticus* in batch and continuous process. *Process Biochem.* 25: 43-46.

Drumright, R.E., Gruber, P.R., Henton, D.E. (2000). Polylactic acid technology. *Adv. Mater.* 12: 1841–1846.

Ejiofor, A.O., Chisti, Y. and Moo Young, M. (1996). Culture of *Saccharomyces cerevisiae* on hydrolyzed waste cassava starch for production of baking quality yeast. *Enz. Microb. Technol.* 18(7): 519- 525.

FAO (2009). Food Outlook: Global Market Analysis on Cassava. http://www.fao.org/docrep/012/ak341e/ak341e06.htm.

Fischer, A., Oehm, C., Selle, M. and Werner, P. (2005). Biotic and abiotic transformations of methyl tertiary butyl ether (MTBE). *Environ. Sci. Pollut. Res. Int.* 12: 381-386.

Fontana, J.D., Passos, M., Baron, M., Mendes, S.V. and Ramos, L.P. (2001). Cassava starch maltodextrinization/monomerization through thermopressurized aqueous phosphoric acid hydrolysis. *Appl. Biochem. Biotechnol.* 91: 449- 480.

Gajria, A.M., Dave, V., Gross R.A. and McCarthy, S.P. (1996). Miscibility and biodegradability of blends of poly (lactic acid) and poly (vinyl acetate). *Polymer* 37: 437-444.

Ghofar, A., Ogawa, S. and Kokugan, T. (2005). Production of L-lactic acid from fresh cassava roots slurried with tofu liquid waste by *Streptococcus bovis* . *J. Biosci. Bioeng.* 100: 606-612

Gutierrez-Correa, M. and Villena, G.K. (2003). Surface adhesion fermentation: A new fermentation category. *Rev. Peru Biol.* 10: 113- 124.

Jacques, K., Lyons, T. P. and Kelsall, D. R. (Eds.) (1999). *The Alcohol Textbook*. Nottingham University Press, UK, 3rd Edition, pp. 388.

Jaleel S. A., Srikanta, S., Ghildyal, N. P. and Lonsane, B. K. (1988). Simultaneous solid phase fermentation and saccharification of cassava fibrous residue for production of ethanol. *Starch - Stärke.* 40: 55 – 58.

Jiang, L., Rozelle, S. and Huang, J. (1993). Production, technology and post harvest processing of sweet potatoes in Sichuan Province, Unpublished Report of CIP Sub-project 6211, pp. 63.

John RP, Nampoothiri KM and Pandey A. (2007). Production of L(+) lactic acid from cassava starch hydrolyzate by immobilized *Lactobacillus delbrueckii. J. Basic Microbiol.* 47(1):25-30.

John, R.P., Madhavan, N.N. and Panday, A. (2006). Solid-state fermentation for L-lactic acid production from agro wastes using *Lactobacillus delbrueckii. Process Biochem.* 41: 759-763.

Johnson R., Padmaja G. and Moorthy, S.N. (2009). Comparative production of glucose and high fructose syrup from cassava and sweet potato roots by direct conversion techniques. *Innov. Food Sci. Emerg. Technol.*10: 616-620.

Jyothi, AN, Sasikiran, K, Nambisan, B, Balagopalan, C. (2005). Optimization of glutamic acid production from starch factory residue using *Brevibacterium divaricatum. Process Biochem.* 40: 3576-3579.

Kim, C.H. and Rhee, S.K. (1993). Process development for simultaneous starch saccharification and ethanol fermentation by *Zymomonas mobilis. Process Biochem.* 28: 331-339.

Kim, C.H., Ryu, Y.W., Kim, C. and Rhee, S.K. (1990). Semi- batch ethanol production from starch by simultaneous saccharification and fermentation using cell recycle. *Kor. J. Biotechnol. Bioeng.* 5: 335-339.

Kim, C.H., Zainal, A., Chong, C.N. and Rhee, S.K. (1992). Pilot scale ethanol fermentation by *Zymomonas mobilis* from simultaneously saccharified sago starch. *Biores. Technol.* 40: 1-6.

Kunhi, A. A. M., Ghildyal, N. P., Lonsane, B. K., Ahmed, S. Y. and Natarajan, C. P. (1981). Studies on production of alcohol from saccharified waste residue from cassava starch processing industries. *Starch - Stärke* 33: 275 – 279.

Laopaiboon, L., Thanonkeo, P., Jaisil, P. and Laopaiboon, P. (2007). Ethanol production from sweet sorghum juice in batch and fed-batch fermentations by *Saccharomyces cerevisiae. World J. Microbiol. Biotechnol.* 23: 1497-1501.

Litchfield, J.H. (1996). Microbiological production of lactic acid. *Adv. Appl. Microbiol.* 42: 45–95.

Lunt, J. (1998). Large-scale production, properties and commercial applications of polylactic acid polymers. *Polym. Degrad. Stabil.* 59: 145–152.

Maddox, I.S., Qureshi, N. and Roberts-Thomson, K. (1995). Production of acetone-butanolethanol from concentrated substrate using *Clostridium acetobutylicum* in an integrated fermentation-product removal process. *Process Biochem.* 30: 209-215.

Naranong, N., Poocharoenm D. (2001). Production of L-lactic acid from raw cassava starch by *Rhizopus oryzae* NRRL 395. www.thaiscience. info/.../Ts2%20production%20of%20llactic%20acid%20from%20raw%20cassava%20starch%20by%20rhi.

Nellaiah, H. and Gunasekaran, P. (1992). Ethanol Production from cassava starch hydrolysate by immobilized *Zymomonas mobilis. Indian J. Microbiol.* 32 (4) : 435-442.

Nijienhuis, A.J., Colstee, E., Grijpma, D.W. and Pennings, A.J. (1996). High molecular weight poly (L-lactide) and poly (ethylene oxide) blends: thermal characterization and physical properties, *Polymer* 37: 5849-5857.

Nuenz, M.J. and Lema, J.M. (1987). Cell immobilization: Application to alcoholic production. *Enz. Microb. Technol.* 9: 642-650.

Ogata, N., Jimenez, G., Kawai H. and Ogihara, T.J. (1997). Structure and thermal/mechanical properties of poly (l-lactide)-clay blend. *J. Polymer Sci. Part B: Poly Phys.* 35: 389- 396.

Panday, A., Soccal, C.R., Nigam, P. and Soccol, V.T. (2000). Biotechnological potential of agroindustrial residues: II Cassava bagasse. *Biores. Technol.* 74: 81-87.

Panesar, P.S., Marwaha, S.S., Gill, S.S. and Rai, R. (2001). Screening of *Zymomonas mobilis* strains for ethanol production from molasses. *Indian J. Microbiol.* 41 : 187-189.

Paolucci, J.D., Belleville, M.P., Zakhia, N. and Rios, G.M. (2000). Kinetics of cassava starch hydrolysis with Termamyl enzyme. *Biotechnol. Bioeng.* 6(1): 71-77.

Pranamuda, H., Lee, S.-W., Ozawa, T. and Tanaka, H. (1994). Ethanol production from raw sago starch under unsterile conditions. *Starch/Starke.* 46: 277-280.

Pradol, F. C., Vandenberghe, L. P. S., Woiciechowski1, A. L., Rodrígues-León, J. A. , Soccol, C. R. (2005).Citric acid production by solid-state fermentation on a semi-pilot scale using different percentages of treated cassava bagasse. *Braz. J. Chem.* 22: 547 – 555.

Qi, F. and Hanna, M.A. (1999) Rheological properties of amorphous and semicrystalline polylactic acid polymers. *Ind. Crops Prod.* 10: 47-53.

Raja, K.C.M., Sreedharan, V.P., Prema, P. and Ramakrishna S.V.(1990). Cyclodextrins from cassava (*Manihot esculenta* Crantz) starch. *Starch/Starke* 42: 196-198.

Rattanachomsri, U., Tanapongpipat, S., Eurwilaichitr, L. and Champreda, V. (2009). Simultaneous non-thermal saccharification of cassava pulp by multi-enzyme activity and ethanol fermentation by *Candida tropicalis. J. Biosci. Bioeng.* 107: 488-493.

Ray, R. C., Mohapatra, S., Panda, S. and Kar, S. (2008). Solid substrate fermentation of cassava fibrous residue for production of α- amylase, lactic acid and ethanol. *J. Environ. Biol.* 29 :111-115.

Ray, R. C., Sharma, P. and Panda, S. H. (2009). Lactic acid production from cassava fibrous residue using *Lactobacillus plantarum* MTCC 1407. *J. Environ. Biol.* 30: 847-52.

Ray, R.C. and Edison, S. (2005). Microbial biotechnology in agriculture and aquaculture- an overview. In: *Microbial Biotechnology in Agriculture and Aquaculture, Volume1* (R.C. Ray, Ed.), Science Publishers, New Hampshire, USA, pp.1- 28.

Ray, R.C. and Moorthy, S.N. (2007). Exopolysaccharide production from cassava starch residue by *Aureobasidium pullulans* strain MTCC 1991. *J. Sci. Ind.Res., India* 66 : 252-255

Ray, R.C. and Naskar, S.K. (2008). Bio-ethanol production from sweet potato (*Ipomoea batatas* L.) by enzymatic liquefaction and simultaneous saccharification and fermentation. *Dyn. Biotechnol. Process Biochem. Mol. Biol.* 2: 47- 49.

Ray, R.C. and Ward, O.P. (2006). Post- harvest microbial biotechnology of topical root and tuber crops. In: *Microbial Biotechnology in Horticulture, Volume 1* (R. C. Ray and O.P. Ward, Eds.), Science Publishers, New Hampshire, USA, pp. 345- 396.

Ray, R.C. Bari, .M. L. and Isobe, S. (2010). Agro – industrial bio-processing of tropical root and tuber crops: Current research and future prospects. In: *Industrial Exploitation of Microorganisms* (D.K. Maheshwari, R.C. Dubey and R.S. Saravanamuthu, Eds.,), I.K. International Publishers House (P) Ltd., New Delhi, India, pp.352- 374.

Ray, R.C., Kar, S. and Chaudhury, P. (2004b). Comparative study of bioethanol production from sweet potato (*Ipomoea batatas* (L) Lam.) and cassava (*Manihot esculents* Crantz) by acid- enzyme hydrolysis. In: National Seminar on Root and Tuber Crops for Nutrition, Food Security and Sustainable Environment, 29-31 October, 2004, Regional Centre of Central Tuber Crops Research Institute, Bhubaneswar, India, Abstract, pp. 61-62.

Ray, S.S. and Bousmina, M. (2005). Biodegradable polymers and their layered silicate nanocomposites in greening the 21st century materials world. *Prog. Mater. Sci.* 50: 962-1079.

Roble, N.D., Ogbonna, J.C. and Tanaka, H. (2003). L-lactic acid production from raw cassava starch in a circulating loop bioreactor with cell immobilized in loofa (*Luffa cylindrica*), *Biotechnol. Lett.* 25: 1093–1098.

Roukas, T. (1998). Pretreatment of beet molasses to increase pullulan production. *Process Biochem.* 33 (8): 805-810.

Selbmann, L., Crognale, S. and Petruccioli, M. (2002). Exopolysaccharide production from *Sclerotium glucanicum* NRRL 3006 and *Botryosphaeria rhodina* DABAC-P 82 on raw and hydrolyzed starchy materials. *Lett. Appl. Microbiol.* 34: 51-55.

Shanavas, S., Padmaja, G., Moorthy, S.N., Sajeev, M.S. and Sheriff, J.T. (2011).Process optimization for bioethanol production from cassava starch using novel eco-friendly enzymes. *Biomass Bioenergy.* 35: 901-909.

Sherman, A. (2000). Ethanol production plans gain power in North Carolina. *Biocycle* 4198: 71-72.

Sheth, M., Kumar, R.A., Dave, V., Gross R.A. and McCarthy, S.P. (1997). Biodegradable polymer blends of poly (lactic acid) and poly (ethylene glycol). *J. Appl. Polym. Sci.* 66: 1495-1505.

Soccol, C. R. and Vandenberghe, L. P. S. (2003). Overview of solid-statefermentation in Brazil. *Biochem. Eng. J.* 13, 205-219.

Södergård, A. and Stolt, M. (2002). Properties of lactic acid based polymers and their correlation with composition, *Prog. Polym. Sci.* 27: 1123–1163.

Sosa, A.V. (2001). Fluidized bed design parameters affecting novel lactic acid downstream processing. *Biotechnol. Prog.* 17: 1079- 1083.

Srikanta, S., Jaleel, S. A., Ghildyal, N. P., Lonsane, B. K. and Karanth, N. G. (1987). Novel technique for saccharification of cassava fibrous waste for alcohol production. *Starch - Stärke.* 39: 234 – 237.

Srinivas, T. and Anantharaman, M. (2005). *Cassava Marketing System in India*. Central Tuber Crops Research Institute, Thiruananthapuram, India. pp. 102.

Sriroth, K., Chollakup, R., Choineeranat, S., Piyachomkwan, K. and Oates, C.G. (2000). Processing of cassava wastes for improved biomass utilization. *Biores. Technol.* 71(1): 63-70.

Tengerdy, R. P. and Szakacs, G. (2003). Bioconversion of lignocellulose in solid substrate fermentation. *Biochem. Eng. J.* 13:169-179.

Thongchul, N., Navankasattusas S. and Yang, S.T. (2010). Production of lactic acid and ethanol by *Rhizopus oryzae* integrated with cassava pulp hydrolysis. *Biopro. Biosyst. Eng.* 33: 407-416.

Toyin, B.O. (2000). Ethanol industry and major technical development in production of ethanol from cassava in Nigeria. In: *Potential of Root Crops*

32   Ramesh C. Ray and Manas R. Swain

for Food and Industrial Resources (M. Nakatani and K. Komaki, Eds.) Twelfth Symposium of International Society of Tropical Root Crops (ISTRC), 10-16 Sept., 2000, Tsukuba, Japan, pp. 121-125.

Tsuji, H. and Ikada, Y. (1996) Blends of isotactic and atactic poly (lactide)s : 2. Molecular-weight effects of atactic component on crystallization and morphology of equimolar blends from the melt. *Polymer* 37: 595-602.

Ueda, S., Zenin, C. T., Monteiro, D. A. and Park Y. K. (1981). Production of ethanol from raw cassava starch by a nonconventional fermentation method. *Biotech. Bioeng.* 23: 291 – 299.

Varadarajan, S. and Miller, D.J. (1999). Catalytic upgrading of fermentation-derived organic acids. *Biotechnol. Progr.* 15: 845–854.

Vink, E.T.H., Rábago, K.R., Glassner, D.A. and Gruber, P.R. (2003). Applications of life cycle assessment to Nature WorksTM polylactide (PLA) production, *Polym. Degrad. Stabil.* 80: 403–419.

Ward, O.P. and Singh, A. (2005). Microbial technology for bioethanol production from agricultural and forestry wastes. In: *Microbial Biotechnology in Agriculture and Aquaculture, Volume I* (R.C. Ray, Ed.), Science Publishers, Enfield, New Hampshire, USA, 449- 479.

Ward, O.P., Singh, A. and Ray, R.C. (2006) Production of renewable energy from agricultural and horticultural substrates and wastes. In: *Microbial Biotechnology in Horticulture, Volume 1* (R. C. Ray and O.P. Ward, Eds.), Science Publishers, New Hampshire, USA, pp. 517- 558.

Wee, Y.J., Kim, J.N. and Ryu, H.W. (2006). Biotechnological production of lactic acid and its recent applications. *Food Technol. Biotechnol.* 44:163–172.

Vandenberghe, L. P. S., Soccol, C. R., Pandey, A. and Lebeault, J.-M. (2000). Solid-state Fermentation for the Synthesis of Citric Acid by *Aspergillus niger. Biores. Technol.* 74: 175-178.

Yu, B., Zhang, F., Zheng, Y. and Wang, P. (1996). Alcohol fermentation from the mash of dried sweet potato with its dregs using immobilized yeast. *Process Biochem.* 31 (1): 1-6.

Zanin, G.M. and de Moraes, F.F. (1998). Thermal stability and energy of deactivation of free and immobilized amyloglucosidase in the saccharification of liquefied cassava starch. *Appl. Biochem. Biotechnol.* 70-72: 383- 394.

In: Cassava: Farming, Uses, and Economic Impact ISBN:978-1-61209-655-1
Editor: Colleen M. Pace © 2012 Nova Science Publishers, Inc.

Chapter 2

POTENTIAL USES OF CASSAVA WASTEWATER IN BIOTECHNOLOGICAL PROCESSES

*Francisco Fábio Cavalcante Barros[1],
Ana Paula Dionísio, Júnio Cota Silva,
and Gláucia Maria Pastore*
Laboratory of Bioflavors. Department of Food Science.
State University of Campinas, Campinas – SP, Brazil

ABSTRACT

Cassava flour is the main cassava derivative for food use in Brazil and its processing generates solid and liquid residues. Concerning the latter, the wash water plus the water extracted from the roots by squeezing is denominated *cassava wastewater* or *manipueira*, and is mainly composed of nutrients such as nitrogen, carbon, potassium, phosphorus, zinc, manganese, calcium, magnesium, sulfur, copper, iron and sodium. This residue is very harmful to the environment due to its high BOD and cyanogenic glycosides. However, the use of this by-product is evident in many areas as an alternative low-cost carbon substrate for use in high-value market compounds. The use of agroindustrial residues in biotechnological processes has been indicated

[1] Corresponding author. Tel./fax: +55 19 3521 4090. E-mail: fabiobarros10@gmail.com. Address: Laboratório de Bioaromas. Departamento de Ciência de Alimentos - FEA, Universidade Estadual de Campinas. Rua Monteiro Lobato, 80. Cx. Postal 6121. 13083-862. Campinas-SP, Brasil.

as an approach to reduce the volume of waste released directly into the environment or involving high costs for effluent treatment. This chapter discusses some applications of this residue in the production of some enzymes (*i.g.* amylases), citric acid, aroma compounds (*i.g.* alcohols), biotransformation processes (*i.g.* culture medium for terpene biotransformation), production of ethanol and, principally, the production of biosurfactants.

INTRODUCTION

Recently, science and industry have joined forces to try and reduce the amount of waste by proposing alternative uses for them. Cassava wastewater (*manipueira*) is an important residue generated in Brazil, responsible for 30% by weight of the raw material in flour mills (Wosiacki *et al.*, 1994). This effluent is usually discharged into the environment, resulting in a major environmental problem, because it contains a large load of organic materials dispersed and in solution. Thus, the high organic load causes a decrease in the oxygen concentration in the water, with damage to the aerobic forms of life (Wosiaki *et al.*, 2000). Fortunately, the environmental problem involving this residue can be changed by the correct use of biotechnological processes. In addition, the reduction in cost has been shown to be a powerful tool to make these processes interesting from an industrial perspective.

An ideal culture medium for use in industrial bioprocesses should provide the nutritional needs for the microorganism, the conditions for product formation and a simpler downstream process, besides presenting low prices (Bicas *et al.*, 2010). According to Stanburry *et al.* (1995), the value of a culture medium represents substantial part of the total production costs depending on the process, and an analysis of the economics of the culture medium has been a key element in making industrial bioprocesses viable and competitive (Bicas *et al.*, 2010). Due to the composition of cassava wastewater (high sugar, nitrogen and mineral contents), this by-product has been considered appropriate for use as a raw material in the production of industrially relevant and value-added compounds (Orzua *et al.*, 2009). In order to elucidate some applications for this residue, this paper reviews the potential uses of cassava wastewater in biotechnological processes.

COMPOSITION

Cassava wastewater is a high carbon source substrate originating from pressing of the roots in the manufacture of cassava flour and starch (Ponte et al., 1992). Quantitatively it is the most important liquid residue from the processing of cassava and presents a milky aspect with a light yellow color. It is produced in abundance, representing about 30% by weight of the raw material in cassava flour factories (Wosiacki et al., 1994, Ponte et al., 2001).

Since they are soluble in water, linamarin and its metabolites are extracted in the wastewater (Cereda, 2001), together with about 5 to 7% starch, 2 to 3% carbohydrates, 1 to 1.5% proteins and less than 1% minerals.

Due to its abundance in all the regions where cassava is cultivated and industrialized, this effluent is generally discharged into the environment, resulting in a considerable environmental problem, since it contains a high load of organic material dispersed in solution. This organic material reduces the oxygen in the water causing damage to the aerobic forms of life in the water (Wosiaki et al., 2000). As previously reported, cassava wastewater contains linamarin, which, when hydrolyzed by linamarase, produces acetone-cyanohydrin. By enzymatic action (α-hydroxynitrile-liase) or spontaneous decomposition, this compound produces cyanic acid (highly volatile) and cyanates, as well as aldehydes. These cyanates are responsible for the insecticidal, acaricidal and nematocidal actions of the compound, whilst the sulfur, also present in large amounts (about 200 ppm) guarantees its notable fungicidal efficiency. The presence of other substances contributes to this anti-fungal action, such as ketones, aldehydes, cyanalanines, lectins and other toxic proteins, inhibitors of amylases and proteinases that act as complementary active ingredients. In addition, the sulfur has insecticidal-acaricidal action (Ponte, 2001).

Physicochemical Composition of Cassava Wastewater

The physicochemical composition of cassava wastewater is variable, principally with respect to the organic matter, depending on the way the root is processed. All residual starch should be removed before discharge or treatment. Cassava wastewater is characterized by containing the majority of the soluble substances from the tuber plus some insoluble substances in suspension. The chemical composition of the wastewater sustains its potential as a nutrient, considering its richness in potassium, magnesium, phosphorus, calcium, sulfur, iron and micro-nutrients in general. However the presence of cyanate is still seen as a considerable obstacle to the use of this residue as a cassava byproduct (Magalhães, 1998).

Table 1. Physicochemical composition of cassava wastewater

Components	Damasceno, 1998	Cereda, 2001	Nitschke, 2003	Maróstica, 2006
Moisture	-	93.75	-	-
pH	5.5	6.3	5.8	5.3
Micronutrients				
P	83.30	160.84	244,5	368,8
K	895.00	1863,50	3472,60	3641,00
Mg	173.00	405,00	519,00	438,10
Fe	8.00	15,35	7,80	2,72
Cu	0.75	1,15	1,00	1,11
Zn	4.50	4,20	2,80	3,01
Mn	1.50	3,70	1,70	3,46
S	38.00	19,50	154,00	61,35
Ca	184.00	227,50	292,53	236,00
N	1.60	0,49	2,08	1,72

APPLICATION OF CASSAVA WASTEWATER IN BIOTECHNOLOGICAL PROCESSES

Production of Enzymes

The use of microorganisms to increase the protein content of fermented cassava products by fermentation can be a source of enzymes. Two important wastes are generated during the processing of cassava tubers, namely, the cassava peel and the liquid squeezed out of the mash (Oboh, 2005). The peel contains toxic cyanogenic glycosides (Oke, 1968), while the liquid contains a heavy load of microorganisms, lactic acid, lysine (from L. coryneformis), amylase (from Saccharomycees spp) and linamarase (from L. delbruckii) capable of hydrolyzing the glycosides (Raimbault, 1998; Akindahunsi et al. 1999). The resulting products of the fermentation of the cassava peel plus the squeezed out liquid, can be dried and used as animal feed (Okafor, 1998, Oboh and Akindahunsi, 2003; Oboh, 2006). Oboh (2005) found amylase activity in the wastewater from cassava mash fermented with pure strains of

Saccharomyces cerevisae together with Lactobacillus delbruckii and Lactobacillus coryneformis for 3 days, and these amylases were active in a wide range of temperature and pH values.

Citric Acid

Citric acid is generally produced by the submerged fermentation of sucrose or molasses using the filamentous fungus Aspergillus niger (Röhr et al., 1983; Vandenberghe et al., 2000b). Cassava wastewater, a carbohydrate-rich residue generated in large amounts during the processing of cassava flour (Nitschke and Costa, 2007), can be a potential substrate for use in the production of citric acid by submerged fermentation. Leonel and Cereda (1995) studied cassava wastewater as a substrate for the growth of the mold Aspergillus niger to produce citric acid, and there was no significant difference in yield between the synthetic media and the fresh cassava wastewater medium, revealing that substantial research is required to make it possible to use cassava wastewater as a substrate for the biosynthesis of citric acid by A. niger.

Other agro-industrial wastes and by-products have been studied using SSF techniques for their potential use as substrates in the production of citric acid (Vandenberghe et al., 2000a). Due to their cellulosic and starchy nature, many low cost substrates are well adapted to solid-state fermentation and can be used to reduce costs in the production of citric acid, such as apple and grape pomace, cassava bagasse, orange and pineapple waste, coffee husk, soy residue, rice and wheat bran (Hang and Woodams, 1984; Khare et al., 1995; Pandey et al., 2000; Soccol et al., 2003; Prado et al., 2005).

Ethanol

Fermentation processes of any material that contains sugar can derive ethanol. The various raw materials used in the manufacture of ethanol via fermentation are composed of sugars, starches and cellulose materials. Most agricultural biomass containing starch can be used as a potential substrate for the production of ethanol by microbial processes (Lin and Tanaka, 2006). The technology for the production of ethanol from starchy materials involves cooking, liquefaction, saccharification, fermentation and distillation (Atthasampunna et al., 1987). Cassava is a very important and cheap source of carbohydrates, especially due to its high starch content (Amutha and Gunasekara, 2001). Thus it could be a good raw material for the production of ethanol by fermentation.

Before the fermentation of starchy materials, they should be enzymatically hydrolyzed by amylases or, more conventionally, by acid hydrolysis. The enzymatic starch hydrolysates (10% reducing sugar) from cassava were fermented by *Zymomonas mobilis* to produce ethanol, and it was found that the yield in ethanol increased almost 10 times (48.1 g/L) by adding 1% yeast extract (Rhee et al., 1984). Prema et al. (1986) demonstrated a preferential consumption of glucose during the fermentation of cassava starch hydrolysate (CSH, commercially available, prepared by acid hydrolysis) by *Saccharomyces cerevisiae*. The higher value found for ethanol production was 56 g/L using 200 g/L CSH, and the conversion efficiency was to the order of 54.9%. It was reported that a mixed culture of *Endomycopsis fibuligera* and *Zymomonas mobilis* could directly and more efficiently ferment cassava starch to ethanol (88% of conversion efficiency) as compared to the monocultures (66%) (Reddy and Basappa, 1996). Amutha and Gunasekara (2001) co-immobilized cells of Saccharomyces diastaticus and Zymomonas mobilis and produced a high ethanol concentration as compared to immobilized cells of S. diastaticus, during the batch fermentation of liquefied cassava starch. The co-immobilized cells produced 46.7 g/L ethanol from 150 g/L liquefied cassava starch, while immobilized cells of the yeast S. diastaticus produced 37.5 g/L ethanol.

Cassava wastewater is a by-product from the cassava manufacturing process which is rich in carbohydrates. The fermentation of cassava waste water by cells of *Saccharomyces cerevisiae* produced 32.7 g/L ethanol. This value represents 93.7% fermentative process efficiency, based on the production of ethanol as calculated from the stoichiometry of the sugars quantified in the cassava wastewater before fermentation (Camili and Cabello, 2007).

Biosurfactants

One of the main conditions for the selection of a residue as a substrate for fermentative processes is the balance in the nutrients that simultaneously promotes cell growth and accumulation of the bio-product. In the case of the biotechnological production of surfactants, substrates with high carbohydrate or lipid contents attend these needs (Mercade and Manresa, 1994; Makkar et al, 2002).

Considering this aspect, cassava wastewater is a viable alternative as a carbohydrate rich substrate. The production of lipopeptide biosurfactants, especially surfactin, by *Bacillus subtilis* in this residue is widely described in the literature. Various experiments have been carried out on a laboratory scale or in bench fermenters using the strains LB1a, LB2a, LB2b, LB6, LB114, LB115, LB117, LBA, LBB and LB262. (Nitschke, Ferraz and Pastore, 2004); LB5a (Nitschke and Pastore, 2003; Nitschke, Haddad et al 2004; Nitschke, Ferraz and Pastore, 2004; Pastore, Santos and Nitschke, 2003; Nitschke and Pastore, 2006) and ATCC 21332 (Nitschke, Ferraz and Pastore, 2004).The use of the strain LB5a on a pilot scale has also been reported, with recovery from the foam (Barros, Ponezi and Pastore 2008). The production of biosurfactants by *Bacillus subtilis*, as has been pointed out, is associated with the logarithmic growth phase (Cooper et al, 1981; Vater, 1986; Kluge et al., 1988; Kim et al., 1997; Costa, 2005).This stage is highly influenced by the nutritional aspects to which the microorganisms were submitted to, since good nutritional conditions are required for culture growth.

Cassava wastewater possesses elevated total carbohydrate indexes (Nitschke and Pastore, 2003; Barros, Ponezi and Pastore 2008; Cabello and Leonel, 2000), the main low molecular weight carbohydrates being sucrose, fructose and glucose (Costa, 2005), a fact also evidenced by the reducing sugar contents described by Barros et al. (2008). This composition characterizes it as a good substrate for the production of biosurfactants by *Bacillus*, since glucose, fructose and sucrose are the best carbon sources for the synthesis of surfactin (Sandrin et al 1990; Peypoux et al, 1999, Lin, 1996). This use was made evident by Nitschke, 2004, who observed a decrease in the sucrose concentration and increase in the glucose and fructose concentrations at the start of fermentation, attributed to hydrolysis of the sucrose molecule. The sucrose was completely consumed in 48 hours and it was observed that the isolate LB5a consumed the sucrose quicker than the strain 21332. According to the author, the presence of maltose and glucose in the culture medium at a more advanced stage suggested enzymatic starch hydrolysis, although the concentration of soluble starch was only 0.4 g/L. The same was described by Barros et al. (2008), who demonstrated complete exhaustion of the total sugars of cassava wastewater in an experiment in a 40L fermenter at the same time as the concentration of reducing sugars increased, a fact probably resulting from enzymatic hydrolysis. At a later moment, after exhaustion of the total sugars, there was a reduction in the concentration of the reducing sugars. In addition, the behavior of the carbohydrate consumption curve was very similar to that of the production of biosurfactants. In another study, Thompson et al. (2000) reported that *B. subtilis* 21332 expressed an α-amylase that allowed for the use of a starch rich residue from potato to produce biosurfactant.

The nitrogen sources are also important in the production of biosurfactants (Sen, 1997; Gu et al, 2005). They are highly varied, depending on the microorganism used, including as from complex organic sources such as corn steep liquor, peptones, fish flour, urea and autolyzed yeasts, to inorganic sources such as ammonium nitrate, ammonium sulfate and others (Mulligan and Gibbs, 1993). In cassava waste water, the NH_4^+ ions and total nitrogen are found at levels from 0.6 to 2.1 g/L, close to the levels described for synthetic media by Cooper and Goldenberg (1987) of 2.7 g NH_4^+/L and by Sen and Swaminathan (2005) of 1.8 g NH_4^+/L. In addition in cassava wastewater a predominance of NH_4^+ ions over NO_3^- ions can be seen (Nitschke, 2004; Costa, 2005; Barros, Ponezi and Pastore, 2008). Studies have shown a preference for organic nitrogen, ammonium and nitrate, in that order, for the production of surface active agents by *Bacillus subtilis* (Davis et al 1999). Davis et al. (1999) showed that the production of surfactin was strongly influenced by the nitrogen metabolism. These authors suggested a connection between the increase in production and cell growth with a limited nitrate presence.

In addition to the carbohydrates and nitrogen sources cassava waste water is also rich in various nutrients described as important for the production of biosurfacants, since it is known that the biosynthesis is also influenced by the phosphate concentration and the presence of minerals such as Fe, Mn and Ca (Cooper et al., 1981; Sen, 1997). The Mn^{2+} levels can vary from 1.5 to 3.9 mg/L and Fe^{2+} from 2.7 to 15.3 mg/L, ions indicated as having a stimulatory effect on the production of surfactin (Cooper et al. 1981; Sen, 1997; Wei and Chu, 2002; Wei and Chu, 2004). Wei and Chu (1998) reported a significant increase in the production of surfactin when the iron concentration of the medium was increased to 4µM, resulting in levels higher than those attributed to genetically-selected strains. Other nutrients that can affect the production are Zn^{2+} ions (Lin, 1996), found at levels between 2.8 and 4.5 mg/L and phosphate ions (Cooper et al., 1981; Lin, 1996; Mulligan et al., 1989).

In another study cassava wastewater was compared with a synthetic medium consisting of a minimal mineral medium containing mineral salts, ammonium nitrate and glucose (Cooper et al., 1981; Sheppard and Cooper, 1991) and to some other residues such as molasses and milk whey (Nitschke, Ferraz and Pastore, 2004). Considering the final surface tension and the CMD of the culture medium after 72 hours of incubation, the cassava wastewater was the substrate showing the lowest values for surface tension, such behavior being related to the elevated production of biosurfactant (Makkar and Cameotra, 1997; Nitschke, Ferraz and Pastore, 2004). The results indicated cassava wastewater as the most adequate culture medium for the production of biosurfactant, showing production levels compatible with that of strains developed in synthetic culture mediums (Wei et al., 2004; Yeh et al., 2006), and also an elevated efficiency in recovery of the surfactant by collecting the foam as described by Davis et al (2001). Of the studies cited, only that of Costa (2005) reported supplementation of the cassava wastewater with yeast extract, peptone, urea, ammonium nitrate, corn steep liquor and milk whey. However the addition of supplements was discouraged due to data showing an insignificant increase in the quantity of biosurfactant produced and also an increase in the difficulty for recovery and purification.

Protein Biomass

Various residues with adequate carbohydrate levels have been used in the production of microbial protein, including cassava wastewater, which was employed in both solid state and submerged liquid fermentations (Menezes, 2000). In addition other studies have shown protein enrichment by cassava byproducts or residues. Amongst these, the bran has been used in many cases with solid state fermentation. The application of cassava wastewater as such in these cases was small, but nevertheless, for didactic purposes, the subject will be considered.

Aspergillus niger, Gliocladium deliquescens, Trichosporum viride, P. elegans and *Aspergillus oryzae* were grown in non-sterile cassava wastewater without supplementation, for the control of the polluting load. The Chemical Oxygen Demand (COD) was reduced by 50% with a small biomass production. With sterilization, the strains *A. niger* 2, *G. deliquescens* and *A. oryzae* 18 reduced the COD by about 75%, *G. deliquescens* producing the greatest amount of biomass. The same fungus was tested in a 30 liter fermenter, where it showed a further increase in biomass production and a reduction in the COD of 80%. With respect to the protein content of the biomass, the values were higher for *A. oryzae* than for *G. deliquescens* (Lamo and Menezes, 1979).

The production of fungal mass by *A. niger*, *A. oryzae* and *Rhizopus sp.* was also studied in sago and starch processing residues. The products presented protein yields of 7.7 to 18%, the maximum amounts being produced by a strain of *Rhizopus* isolated from a cassava processing location. It was also found that the biomass and protein yields were higher in the residue than in the pure starch medium (Bagalopal and Mani, 1977). In another study, using *Rhodopseudomonas gelatinosa* in a medium containing cassava starch, the temperature, aerobiosis and light conditions were studied, the best results being the production of 0.83 g cells/g starch (Noparatnaraporn et al., 1983). In addition, Essers (1994) reported the potential of *Neurospora sitophila*, *Geotrichum candidum* and *Rhizopus oryzae* in the protein enrichment of cassava products, and Balagopalan (1996) also showed the ability of yeasts to enrich the protein level.

One of the limitations for the use of these residues is that some microorganisms lack the production of enzymes capable of hydrolyzing the starch, and thus some studies were developed concerning this aspect. In one study, the amylolytic yeast *Schwanniomyces alluvius* was cultivated in a non-hydrolyzed cassava substrate, producing a protein content of 31.55% in relation to the dry matter, a value comparable to that produced by *Rhodotorula gracilis* when developed in the same, but hydrolyzed, substrate (Menezes et al., 1971). Subsequently an association was established between the amylolyitic yeasts and *Candida utilis* or *R. gracilis* in order to obtain a better conversion rate of the sugars into protein. The association of *Endomycopsis fibuligera* with *C. utilis* was that presenting the best yield. In a medium containing 30 g/L of total sugars and urea as the source of nitrogen a biomass yield to the order of 14.9% was obtained with a protein content of 56% in the dry matter (Sales and Menezes, 1976). In addition, *Candida tropicalis* CBS 69 48 grew in the non-hydrolyzed cassava and corn starches, since the yeast contained the necessary hydrolytic enzymes, and the protein levels increased from 3.1 to 18.8% (Azoulay et al., 1980).

Some studies evaluated the nutritional profile and processing characteristics of these proteins. In a nutritional evaluation of the use of single cell protein (SCP) from *Aspergillus fumigatus* 1-21A to feed rats and pigs, the biomass resulting from the fermentation reached 34.3%, the nutritive quality of the fungal protein being similar to that obtained from soy bran (Santos and Gómez, 1983). In another study, the batch production of protein by *C. tropicalis* and *C. utilis* from a hydrolyzed cassava substrate supplemented with ammonium hydroxide was reported, reaching protein levels of 71% and 68%, respectively. In addition the protein concentrates showed an amino acid profile comparable to that of egg, and the concentrates presented good processing characteristics with respect to their wettability and emulsifying capacity (Okezie and Kosikowski, 1981).

The production of fungal biomass for animal feed by the submerged fermentation of cassava scrapings and leaves by *A. niger* and *Rhizopus* sp, produced biomass contents above 18.5 g/L with 34% protein and a conversion efficiency of 0.56 and 0.61, respectively. Using semi-solid fermentation, the protein content of the biomass of the *Rhizopus* sp was 9.64% on a dry weight basis. The addition of dehydrated cassava leaves to the substrate at a rate of 19% resulted in an increase in protein content of the fermented biomass to 13.74%, whilst the HCN content was reduced by half (Sales, 1982). In another study the change in protein content of the flour, peel and leaves when submitted to solid state fermentation by *A. niger, Saccharmyces cerevisie, Rhizomucor miehei* and *Mucor strictus* was investigated. All the fungi caused an increase in the protein content, but the best results for the peel were obtained with *A. Niger* and *S. Cerevisiae*, with a 192.85% increase after 21 days, and for the leaves with *R. miehei, M. strictus* and *A. niger* (Iyayi and Losel, 2001). The increase in protein content reported in the present study was similar to that reported by Brook *et al* (1969) using *Rhizopus oligosporus* and *R. stolonifer* in solid state fermentation, by Daubresse et al. (1987), who increased the protein content from 1.0% to 10.7% of the dry matter using *Rhizopus oryzae* strain MUCL 28627, and in the study by Wainright (1992), in which the addition of malt extract or molasses in the initial stage assured an additional growth of the protein content.

More recently studies related to the use of solid residues in the production of glutamic acid by *Brevibacterium divaricatum* MTCC 1529 in a submerged production system, obtained results from which the authors concluded that these residues could be an alternative for the production of this amino acid (Jyothi et al., 2005).

Lipid Biomass

Some reports by a research group have shown cassava wastewater as a substrate for the biotechnological production of an oleaginous biomass by *Trichosporon* sp (Wosiacki, 1994). These studies were carried out to evaluate the growth of this microorganism in a synthetic medium similar to cassava wastewater and determine its basic nutritional demands (Wosiacki et al, 1994). This microorganism was shown to possess the capacity to produce amylolytic enzymes which assured its survival in starch-containing culture media, although the presence of other species also allowed for the liberation of soluble sugars. The production of biomass by *Trichosporon* sp can be carried out using submerged fermentation techniques under controlled conditions (Wosiacki et al., 1995), or more simply by surface fermentation techniques, similar to that occurring naturally in lakes containing industrial residues.

The cultivation conditions for fermentation by *Trichosporon* sp., aimed at obtaining biomass and using the sugars, was studied in a medium containing hydrolyzed starch, HCN, yeast extract and minerals. In addition the surface-volume ratio was also evaluated, since it had been observed that the best conditions occurred as this ratio increased (Efing and Wosiacki, 1998). As a continuation, the same authors established the cultivation conditions for the same microorganism using surface fermentation in a liquid synthetic medium with a composition similar to that of cassava wastewater, which was shown to be efficient in carrying out the fermentation (Efing and Wosiacki, 1999). Subsequently a synthetic medium was formulated that simulated cassava wastewater in a surface fermentation method, determining the influence of the degree of hydrolysis of the starch in the medium for the production of microbial biomass, lipid biomass and the use of the sugars. The best results were obtained with the highest degrees of starch hydrolysis. In addition, the growth profile showed that the strain was dependent on the degree of fragmentation, suggesting that a preliminary hydrolysis procedure was necessary for a complete use of the carbohydrates (Kirchner, Almeida and Wosiacki, 2000). The formation of biomass by *Trichosporon pullulans* in cassava wastewater, supplemented or otherwise with nitrogen, was also investigated. Supplementation was shown to be beneficial, increasing the formation of biomass by 138% as compared to the control, ammonium sulfate being the best source (Almeida et al., 1998).

Heat treatments also show an influence on the microbial flora of cassava wastewater. Thus an investigation was made with fermentations using no heat treatment, sterilization and exposition to temperatures of 60°C and 70°C for 30 minutes. Even the mildest heat treatment was shown to reduce the viable cell count by 99%, resulting in the production of a better quality biomass with no alteration in the sugar consumption (Almeida et al., 1998). Another aspect explored was the development of bench-scale reactors adapted for the continuous admission of culture medium, resulting in the continuous production of biomass, with a 30% decrease in the initial COD load (Wosiacki et al., 1996).

Aroma Compounds

The use of cassava wastewater as a substrate for the production of volatile compounds by *Geotrichum fragans*, a cyanide-resistant microorganism, was evaluated by Damasceno and collaborators (2003). The following volatile compounds were analyzed by CG-FID after 72 hr of fermentation: 1-butanol, 3-methyl 1-butanol (isoamyl alcohol), 2-methyl 1-butanol, 1-3 butanediol, phenylethanol, ethyl acetate, ethyl propionate, 2-methyl ethyl propionate and 2-methylpropenoic acid. Supplementation of the culture medium with glucose or fructose as carbon sources showed no qualitative effect on the profile of the volatile compounds produced by this microorganism.

Biotransformation Process

Maróstica and Pastore (2007) evaluated the use of cassava wastewater in the biotransformation of limonene into R-(+)-α-terpineol, a volatile derivative appreciated in the aroma industry. The agroindustrial residue was used in the development of biomass by *Penicillium* sp. 2025, *Aspergillus* sp. 2038 and *Fusarium oxysporum* 152B. The maximum α-terpineol concentration was obtained using strain 152B, with an initial growth in cassava wastewater and subsequent inoculation into the mineral medium, as compared to the production by direct inoculation into the mineral medium, showing the potential of this agroindustrial residue in the development of biomass.

Other Applications

Cassava wastewater was first studied as a nematocidal agent in 1979 (Ponte *et al.*, 1979). Sacked soils, previously infested with the eggs and larvae of the branch nematode (*Meloidogyne* spp.) were planted with okra plants (*Hibiscusesculentus* L.). After 10 days of cultivation, they were treated with different concentrations of cassava wastewater (0, 500, 750 and 1000 mL per vase), extracted from a mixture of toxic (containing HCN) and non-toxic cassava. The results were highly satisfactory, and showed that the higher the concentration of cassava wastewater applied, the lesser the plants were attacked by Meloidoginose (100, 60, 50 and 30%, respectively). Two years after the first experiment, Ponte and Franco (1981) obtained even better results using the cassava wastewater extracted exclusively from toxic cassava. Sena and Ponte (1982) corroborated these results in the control of Meloidoginose in carrot (*Daucus carota* L.) plantations. Many other papers have been published concerning the nematocidal potential of cassava wastewater, aiming to determine the appropriate dosage and storage time, amongst other factors.

In tests carried out by Ponte et al. (1988), they added pure cassava wastewater to Galego lemon (*Citrus aurantifolia* Swingle) stock plagued with brown-shelled cochineal insects (*Coccus hesperidum* L.), and showed the potential of this residue as an insecticidal agent. This was followed by various other tests in which they reported the beneficial effects of cassava wastewater with distinct agricultural pests.

With the diffusion of positive results for research carried out using cassava wastewater as an insecticide, its use has become a routine practice in various points of the Brazilian territory, as also in other parts of the World, such as Madagascar, and Africa, amongst others (Ponte et al., 2001). In addition to the use of cassava wastewater as a nematocidal and insecticidal agent, some reports suggest its use as a substrate in the production of oleaginous biomass by *Trichosporon sp* (Wosiacki, 2000).

REFERENCES

Almeida, M. M.; Kanunfre, C. C.; Kirchner, C. L.; Wosiacki, G. (1998) Produção de biomassa de *Trichosporon* sp. Fermentação em manipucira utilizando diferentes fontes de nitrogênio. *Bol. CEPPA*, 16(2), 247-262.

Amutha, R. and Gunasekara, P. (2001) Production of ethanol from liquefied cassava starch using co-immobilized cells of *Zymomonas mobilis* and *Saccharomyces diastaticus*. *Journal of Bioscience and Bioengineering*, 92(6), 560-564.

Atthasampunna, P.; Somchai, P.; Eur-aree, A.; Artjariyasripong, S. (1987) Production of fuel ethanol from cassava. *World Journal of Microbiology and Biotechnology*, 3(2), 135-142.

Azoulay, E.; Jouanneau, F.; Bertrand, J-C.; Raphael, A.; Janssens, J.; Lebeault, J. M. (1980) Fermentation methods for protein enrichment of cassava and corn with *Candida tropicalis*. *Applied and Environmental Microbiology*, 39(1), 41-47.

Balagopal, C. Maini, S. B. (1997) Studies on the utilization of cassava waste for single cell protein. *Journal of Root Crops*, 3(1), 33-36.

Balagopalam, C. (1996). Nutritional improvement of cassava products using Microbial techniques, for animal feeding. Central Tuber Crops Research Publication, 1996. Monograph of the Central Tuber Crops Research Institute, Kerala, India.

Barros, F. F. C.; Ponezi, A. N.; Pastore, G. M. (2008) Production of biosurfactant by *Bacillus subtilis* LB5a on a pilot scale using cassava wastewater as substrate, *Journal of Industrial Microbiology and Biotechnology*, 35, 1071-1078.

Bicas, J. L. ; Dionísio, A.P. ; Silva, J. C.; Barros, F. F. C. B.; Pastore, G. M. (2009) Agro-Industrial Residues in Biotechnological Processes. In: Jürgen Krause; Oswald Fleischer. eds., Industrial Fermentation: Food Processes, Nutrient Sources and Production Strategies. Hauppauge: Nova Publishers.

Brook, E. J., Stanton W. R., Wallbridge A. (1969) Fermentation methods for protein enrichment of cassava by solid substrate fermentation in rural conditions fermentation. *Acta Horticultural*, 375, 217-224.

Cabello, C; Leonel. M. (2000) Produção de ácido cítrico a partir do resíduo líquido da industrialização da mandioca (manipueira). In: Cereda, M. P., ed., Manejo, uso e tratamento de subprodutos da industrialização da mandioca (chap. 9); São Paulo: Fund. Cargill.

Camili, E. A.; Cabello, C. (2007) Produção de etanol de manipueira tratada com processo de flotação. Revista Raízes e Amidos Tropicais, 3.

Cassoni, V. (2008) Valorização de resíduo de processamento de farinha de mandioca (manipueira) por acetificação (thesis). Botucatu: Universidade estadual paulista "Julio de Mesquita Filho" Faculdade de Ciências Agronômicas.

Cereda, M. P. (2001) Caracterização dos subprodutos da Industrialização da Mandioca. In: Cereda M. P. ed., Manejo, Uso e tratamento de subprodutos da Industrialização da mandioca (chap 1). São Paulo: Fundação cargill, 13-37.

Cereda, M. P. (2005) Fécula de mandioca como ingrediente para alimentos. Revista da ABAM, n.11.

Cooper, D. G.; Macdonald, C.R.; Duff, S.J.B.; Kosaric, N. (1981) Enhanced production of surfactin from *Bacillus subtilis* by continuous product removal and metal cation additions. *Applied Environmental Microbiology*, 42, 408-412.

Costa, G. A. N. (2005) Produção biotecnológica de surfactante de *Bacillus subtilis* em resíduo agroindustrial, caracterização e aplicações (thesis). Campinas: Universidade Estadual de Campinas.

48   Francisco Fábio Cavalcante Barros, Ana Paula Dionísio et al.

Damasceno, S. (1998) Manipueira como substrato para desenvolvimento de *Geotrichum fragans* (thesis) Botucatu: Faculdade de Ciências Agronômicas, Universidade Estadual Paulista.

Damasceno, S.; Cereda, M. P.; Pastore, G. M.; Oliveira, J. G. (2003) Production of volatile compounds by *Geotrichum fragans* using cassava wastewater as substrate. *Process Biochemistry*, 39, 411-414.

Daubresse, P. S.; Nitbashirwa, S.; Gheyen, A.; Meyer, J. A. (1987) A process for protein enrichment of cassava by solid substrate fermentation in rural conditions. *Biotecthnology Bioengineering*, 29, 962-968.

Davis, D. A.; Lynch, H. C.; Varley, J. (2001) The application of foaming for the recovery of surfactin from *Bacillus subtilis* ATCC 21332 cultures. *Enzyme and Microbial Technology*, 28, 346-354.

Efing, L. de M. A. C.; Wosiacki, G. (1998) Estabelecimento de condições de cultivo de uma cepa de *Trichosporon* sp isolada de manipueira. Boletim do Centro de Pesquisa e Processamento de Alimentos (Brazil). 16(1), 23-36.

Essers, A. J. (1994) Making safe flour from bitter cassava by indigenous solid substrate fermentation. *Acta Horticultural*, 375, 217-224.

FAOSTAT (2006) Faostat database. www.faostat.org.

Fernandes Jr., A. (2001) Tratamentos físicos e biológicos da manipueira. In: Cereda, M.P., ed. Manejo, uso e tratamento de subprodutos da industrialização da mandioca (chap 10). São Paulo: Fundação Cargill, 138-50.

Gu, X.; Zheng, Z.; Yu, H.; Wang, J.; Liang, F.; Liu, R. (2005) Optimization of medium constituents for a novel lipopeptide production by *Bacillus subtilis* MO-01 by a response surface method. *Process Biochemistry*, 40, 3196-3201.

Hang, Y. D.; Woodams, E. E. (1984) Apple Pomace: A Potential Substrate for Citric Acid Production by *Aspergillus niger*, *Biotechnology Letters*, 6, 763-764.

Iyayi, E. A.; Losel, D. M. (2001) Changes in carbohydrate fractions of cassava peel following fungal solid state fermentation. *The Journal of Food Technology in Africa*, 6, 101-103.

Jyothi A. N., Sasikiran K., Nambisan Bala, Balagopalan C. (2005) Optimisation of glutamic acid production from cassava starch factory residues using Brevibacterium divaricatum. *Process Biochemistry*, 40, 3576–3579.

Khare, S. K., Jha, K. and Gandhi, A. P. (1995) Short Communication: Citric Acid Production from Okara (Soy-residue) by Solid-state Fermentation, *Bioresource Technology*, 54, 323-325.

Kim, H.; Yoon, B.; Lee, C.; Suh, H.; Oh, H.; Katsuragi, T.; Tani, Y. (1997) Production and properties of a lipopeptide biosurfactant from *Bacillus subtilis* C9. *Journal of Fermentation and Bioengineering*, 84, 41-46.

Kirchner, C. L.; Almeida, M. M.; Wosiacki, G. (2000) Production of Trichosporon sp biomass – Surface fermentation with dextrin as source of carbon. UEPG – Ciências Exatas e da Terra, Agrárias e Engenharias, 6.

Kluge, B.; Vater, J.; Salnikow, J.; Eckart, K. (1988) Studies on the biosynthesis of surfactin, a lipopeptide antibiotic from *Bacillus subtilis* ATCC 21332. *FEBS Letters*. 231, 107-110.

Lamo, P. R., Menezes, T. J. B. (1979) Bioconversão da águas residuais do processamento de mandioca para produção de biomassa. *Col ITAL*, 10, 1-14.

Leonel, M.; Cereda, M. P. (1995) Citric acid production by *Aspergillus niger* from "manipueira", a manioc liquid residue. *Scientia Agricola*, 52, 299-304.

Lin, S.C. (1996) Biosurfactants: recent advances. *Journal of Chemical Technology and Biotechnology*. 66: 109-120, 1996.

Lin, Y. and Tanaka, S. (2006) Ethanol fermentation from biomass resources: current state and prospects. *Applied Microbiology and Biotechnology*, 69, 627-642.

Magalhães, C. P. (1998) Estudos sobre as bases bioquímicas da toxicidade da manipueira a insetos, nematóides e fungos (thesis). Fortaleza: Universidade Federal do Ceará.

Makkar, R. S.; Cameotra, S. S. (1997) Utilization of molasses for biosurfactant production by two *Bacillus* strains at thermophilic conditions. *Journal of American Oil Chemist's Society*, 74 (7), 887-889.

Makkar, R. S.; Cameotra, S. S. (2002) An update on the use of unconventional substrates for biosurfactant production and their new applications. *Applied Microbiology and Biotechnology*. 58, 428-434.

Maróstica Jr, M. R. (2006) Biotransformação de terpenos para a produção de compostos de aroma e funcionais (thesis). Campinas: Faculdade de Engenharia de Alimentos, Universidade Estadual de Campinas.

Maróstica Jr, M. R.; Pastore, G. M. (2007) Production of R-(+)-α-terpineol by the biotransformation of limonene from orange essential oil, using cassava waste water as medium. *Food Chemistry*, 101, 345-350.

Menezes, T. J. B. (2000) Produção de biomassa protéica a partir da manipueira. In: Cereda, M. P., ed., Em Manejo, uso e tratamento de subprodutos da industrialização da mandioca (chap 8); São Paulo: Fundação Cargill: São Paulo.

Menezes, T. J. B., Figueiredo, I. B., Strasser, J. (1971) Proteína monocelular de leveduras amilolíticas, *Col. ITAL*, 4, 109-116.

Mercade, M. E.; Manresa, M. A. (1994) The use of agroindustrial by-products for biosurfactant production. *Journal of American Oil Chemist's Society*, 71 (1): 61-64.

Mulligan, C. N.; Gibbs, B. F. (1993) Factors influencing the economics of biosurfactants. In: Kosaric, N. ed., Biosurfactants: production, properties, applications. New York, Marcel Decker Inc., 392-371.

Mulligan, C. N.; Mahmourides, G.; Gibbs, B.F. (1989) The influence of phosphate metabolism on biosurfactant production by *Pseudomonas aeruginosa*. *Journal of Biotechnology*, 12, 199-210.

Nitschke, M. and Costa, S. G. V. A. O. (2007) Biosurfactants in food industry. *Trends in Food Scidnce and Technology*, 18, 252-259.

Nitschke, M. (2004) Produção e caracterização de biossurfactante de *Bacillus subtilis* utilizando manipueira como substrato. (thesis). Campinas: Universidade Estadual de Campinas.

Nitschke, M. Pastore, G.M. (2003) Cassava flour wastewater as a substrate for biosurfactant production. *Applied Biochemistry and Biotechnology*. 106, 295-302.

Nitschke, M.; Ferraz, C.; Pastore, G. M. (2004) Selection of microrganisms for biosurfactant production using agroindustrial wastes. *Brazilian Journal of Microbiology*. 35, 81-85.

Nitschke, M.; Haddad, R.; Costa, G. A. N.; Gilioli, R.; Meurer, E. C.; Gatti, M. S.; Eberlin, M. N.; Höehs, N. F.; Pastore, G. M. (2004) Strutural characterization and biological properties of a lipopeptide surfactant produced by *Bacillus subtilis* on cassava wastewater medium. *Food Science and Biotechnology*, 13, 591-596.

Nitschke, M; Pastore, G. M. (2006) Production and properties of a surfactant obtained from *Bacillus subtilis* grown on cassava wastewater. *Bioresourse Technology*, 97, 336-341.

Noparatnaraporn, N.; Nishizawa, Y.; Hayashi, M.; Nagai, S. (1981) Single cell protein production from cassava starch by *Rhodopseudomonas gelatinosa*. *Journal of Fermentation Technology*, 61(5), 515-519.

Oboh, G. (2005) Isolation and characterization of amylase from fermented cassava (*Manihot esculenta* Crantz). *African Journal of Biotechnology*, 4, 1117-1123.

Oboh, G. (2006) Nutrient enrichment of Cassava peels using a mixed culture of *Saccharomyces cerevisae* and *Lactobacillus spp* solid media fermentation techniques. *Electronic Journal of Biotechnology*, 9, 46-49.

Oboh, G.; Akindahunsi, A. A. (2003) Chemical changes in cassava peels fermented with mixed culture of *Aspergillus niger* and two species of *Lactobacillus* integrated bio-system. *Applied Tropical Agriculture*, 8 63–68.

Okafor, N. (1998) An integrated bio-system for the disposal of cassava wastes, integrated bio-systems in zero emissions applications. Proceedings of the internet Conference on integrated Bio-Systems.

Oke, O.L. (1968) Cassava as food in Nigeria. *World Review of Nutrition and Dietetics*, 9, 227-250.

Okezie, B. O., Kosikowski F. V. (1981) Extractability and functionality of protein from yeast cells grown on cassava hydrolysate. *Food Chemistry*, 7(1), 7-18.

Orzua, M. C.; Mussatto, S. I.; Contreras-Esquivel, J. C.; Rodriguez, R.; de la Garza, H.; Teixeira, J. A.; Aguilar, C. N. (2009) Exploitation of agro industrial wastes as immobilization carrier for solid-state fermentation. *Industrial Crops and Products*, 30, 24-27.

Pandey, A., Soccol, C. R., Nigam, P., Soccol, V. T. and Mohan R. (2000) Biotechnological Potential of Agro-industrial Residues. II. *Cassava Bagasse, Bioresource Technology*, 74, 81-87.

Pantoratoro, S. (2001) Isolamento, seleção, identificação e avaliação de microrganismos anaeróbios "in situ", com habilidade à biodegradação de linamarina (thesis). Botucatu: Faculdade de Ciências Agronômicas, Universidade Estadual Paulista.

Peypoux, F.; Bonmartin, J. M.; Wallach, J. (1999) Recent trends in the biochemistry of surfactin. *Applied Microbiology and Biotechnology*, 51, 553-563.

Ponte J. J.; Franco A.; Santos J. H. R. (1992) Eficiencia da manipueira no controle de duas pragas da citricultura. In: Congresso Brasileiro de mandioca, 7., Recife, Anais. Recife: Sociedade Brasileira de mandioca. p. 59.

Ponte, J. J. (1981) Franco, A. Manipueira, um nematicida não convencional de comprovada potencialidade. Public. *Sociedade Brasileira de Nematologia, Piracicaba*, 5, 25-33.

Ponte, J. J. (1992) Histórico das pesquisas sobre a utilização da manipueira (extrato líquido das raízes da mandioca) como defensivo agrícola. *Fitopatologia Venezuelana, Maracay*, 5, 2-5.

Ponte, J. J. (2001) Uso da Manipueira como Insumo Agrícula: Defensivo e Fertilizante. In: Cereda M. P., ed., Manejo, Uso e tratamento de subprodutos da Industrialização da mandioca (chap 5). São Paulo: Fundação Cargill.

Ponte, J. J.; Franco, A.; Santos, A. E. L. (1988) Teste preliminar sobre a utilização da manipueira como inseticida. Ver. Bras. Mand., *Cruz das Almas,* 7(1), 89-90.

Ponte, J. J.; Torres, J.; Franco, A. (1979) Investigação sobre a possível ação nematicida da manipueira. *Fitopatologia Brasileira, Brasília,* 4, 431-435.

Prado, F. C., Vandenberghe, L. P. S., Woiciechowski, A. L., Rodrígues-León, J. A., Soccol, C. R. (2005) Citric acid production by solid-state fermentation on a semi-pilot scale using different percentages of treated cassava bagasse. *Brazilian Journal of Chemical Engineering*, 22, 547-555.

Prema , P.; Ramakrishna, S. V. and Madhusudhana Rao, J. (1986) Influence of composition of sugars in cassava starch hydrolysate on alcohol production. *Biotechnology Letters*, 8(6), 449-450.

Raimbault, M. (1998) General and microbiological aspects of solid substrate fermentation. *Electronic Journal of Biotechnology*, 1, 174-188.

Reddy, O. V. S. and Basappa, S. P. Direct fermentation of cassava starch to ethanol by mixed cultures of Endomycopsis fibuligera and Zymomonas mobilis: Synergism and limitations. *Biotechnology Letters*, 18(11), 1315-1318, 1996.

Rhee, S. K.; Lee, G. M.; Han, Y. T.; Yusof, Z. A. M.; Han, M. H. and Lee, K. J. Ethanol production from cassava and sago starch using *Zymomonas mobilis. Biotechnology Letters*, 6(9), 615-620, 1984.

Röhr, M., Kubicek, C. P. and Komínek, J. (1983) Citric Acid. In: Reed, G., Rehm, H. J. (Eds.). Biotechnology, vol. 3. Weiheim: *Verlag Chemie*, 419-454.

Sales, A. M. (1982) Produção e avaliação nutricional de biomassa de mandioca fermentada (thesis). São Paulo: Faculdade de Ciências Farmacêuticas, Universidade de São Paulo.

Sales, A. M., Menezes, T. J. B. (1976) Produção de biomassa protéica de mandioca. *Col ITAL*, 7, 139-146.

Sandrin, C.; Peipoux, F.; Michel, G. (1990) Coproduction of surfactin and iturin A lipopeptides with surfactant and antifungal properties by *Bacillus subtilis. Biotechnology and Applied Biochemistry*, 12, 370-375.

Santos J.; Gómez G. (1983) Fungal protein produced on cassava for growing rats and pigs. *Journal of Animal Science*, 56 (2), 264-270.

Sen, R. (1997) Response surface optimization of the critical media components for the production of surfactin. *Journal of Chemical Technology and Biotechnology*. 68, 263-270.

Sena, E. S.; Ponte, J. J. (1982) A manipueira no controle de Meloidoginose da cenoura. Public. Sociedade Brasileira Nematologia, *Piracicaba*, 6, 95-98.

Sheppard, J. D.; Cooper, D. G. (1991) The response of *Bacillus subtilis* ATCC 21332 to manganese during continuous-phased growth. *Applied Microbiology and Biotechnology*, 35, 72-76.

Soccol, C. R. and Vandenberghe, L. P. S. (2003) Overview of applied solid-state fermentation in Brazil, *Biochemical Engineering Journal*, 13, 205-218.

Stanburry, P. F.; Whitaker, A; Hall, S. J. (1995) Principles of Fermentation technology. 2nd edition. Oxford: Butterworth Heinemann.

Théry, H. (2005) Situações da Amazônia no Brasil e no continente. Estudos Avançados. São Paulo. 19 (53), 37-49.

Thompson, D. N.; Fox, S. L.; Bala, G. A. (2000) Biosurfactants from potato process effluents. *Applied Biochemistry and Biotechnology*, 84/86, 917-929.

Vandenberghe, L. P. S., Soccol, C. R., Pandey, A. and Lebeault, J.-M. (2000a) Cassava Bagasse, An Alternative Substrate for Citric Acid Production in Solid-state Fermentation. In: 11th International Biotechnology Symposium and Exhibition, 3-8 Sept. 2000, Berlin. Book of Abstracts. Frankfurt: *Dechema*, vol. 4, 153-155.

Vandenberghe, L. P. S., Soccol, C. R., Pandey, A. and Lebeault, J.-M. (2000b) Solid-state Fermentation for the Synthesis of Citric Acid by Aspergillus niger, *Bioresource Technology*, vol. 74, 175-178.

Vater, J. (1986) Lipopeptides, an attractive class of microbial surfactants. *Progress in Colloid and Polymer Science*, 72, 12-18.

Wainright, M. (1992) An introduction to Fungal Biotechnology. Wiley Biotechnolgy Series. Wiley Publishers. U.K.

Wei, Y. Chu, I. (2002) Mn^{2+} improves surfactin production by *Bacillus subtilis*. *Biotechnology Letters*, 24, 479-482.

Wei, Y. Chu, I. (2004) Optimizing iron supplement strategies for enhanced surfactin production with *Bacillus subtilis*. *Biotechnology Progress*, 20, 979-983.

Wei, Y. H. ; Chu, I. M. (1998) Enhancement of surfactin production in iron-enriched media by *Bacillus subtilis* ATCC 21332. *Enzyme and Microbial Technology*, 22, 724-728.

Wosiacki G, Fioretto AMC, Cereda MP (1994). Utilização da manipueira para produção de biomassa oleaginosa. In: Cereda, M. P., ed., Resíduos da industrialização da mandioca. São Paulo: *Paulicéia*. 151-161.

Wosiacki, G.; Fioretto, A. M. C.; Almeida, M. M.; Cereda, M. P. Utilização da manipueira para produção de biomassa. In: Cereda, M. P., ed., Manejo, uso e tratamento de subprodutos da industrialização da mandioca (chap 12); São Paulo: Fund. Cargill.

Wosiacki, G.; Kirchner, C. L.; Mendes, M. (1993) Produção de enzimas amilolíticas pelo microrganismo *Trichosporon* sp. Curitiba: Evento de Iniciação Científica da UFPR – EVICI, 1.

Wosiacki, G.; Sichieri, V. L. F. S.; Cereda, M. P.; Silva R. S. F.; Bruns, R. E. (1995) Improved submerged fermentation conditions for *Trichosporon* sp. *Arquivos de Biologia e Tecnologia*, 38(2), 405-416.

Wosiacki, G.; Sokoloskia, A.; Duarte, F. (1996) Construção de um protótipo de reator para a produção de biomassa oleaginosa de *Trichosporon* sp. Ponta Grossa: Maratona de pesquisa em Ciência e tecnologia de Alimentos.

Yeh, M.; Wei, Y.; Chang, J. (2006) Bioreactor design for enhanced carrier-assisted surfactin production with *Bacillus subtilis*. *Process Biochemistry*, 41, 1799-1805.

In: Cassava: Farming, Uses, and Economic Impact   ISBN:978-1-61209-655-1
Editor: Colleen M. Pace   © 2012 Nova Science Publishers, Inc.

Chapter 3

BIOTECHNOLOGY APPLIED TO CASSAVA PROPAGATION IN ARGENTINA

M. Cavallero[1], R. Medina[2], R. Hoyos[1],
P. Cenóz[3] and L. Mroginski[2]*
[1]Instituto Nacional de Tecnología Agropecuaria (INTA),
Estación Experimental Agropecuaria (EEA) El Colorado,
Formosa, Argentina
[2]Facultad de Ciencias Agrarias (FCA), Universidad Nacional
del Nordeste (UNNE). Instituto de Botánica del Nordeste – Consejo
Nacional de investigaciones Científicas y Técnicas (CONICET).
Casilla de Correo 209, (3400) Corrientes, Argentina
[3]FCA, UNNE, Argentina

ABSTRACT

Cassava is a staple food to millions of people in tropical and subtropical countries. Although it is traditionally cultivated from stem cuttings, which is a simple and inexpensive technique, this method presents serious problems such as low multiplication rates, difficulties to conserve stems, and dissemination of pests and diseases. Many of these problems would be solved through in vitro tissue culture. This work evaluates the in vitro establishment and multiplication of 28 cassava

* Casilla de Correo 209, (3400) Corrientes, Argentina; Te: +54-3783-427589; Fax: +54-3783-427131. e-mail: ricardomedina@agr.unne.edu.ar.

clones of agronomic interest for the Northeastern Argentina, a boundary area for this crop. Since the transfer of in vitro plants to ex vitro conditions is a critical phase of micropropagation, we evaluated the effect of different acclimatization treatments on survival and growth parameters of plants (cv EC118) grown in a culture chamber. We also scored their field survival and performance by comparing them with plants obtained by the conventional planting technique. After disinfection, uninodal segment culture in Murashige and Skoog medium supplemented with 0.01 mg/L BAP + 0.01 mg/L NAA + 0.1 mg/L GA3 allowed the in vitro establishment of 100% of the clones and their subsequent multiplication. Cultures were maintained at 27°±2°C with a 14 h photoperiod. During establishment, sprouting occurred in 100% of the clones and rooting in 93% of them; the remaining clones formed roots during the multiplication phase. Thirty days after multiplication, the plants presented significant differences in plant height, average number of nodes per plant and number of roots per plant. During acclimatization, five treatments were evaluated: three substrates (perlite, T1; sand + vermicompost, T2; commercial substrate composed of peat and perlite, T3), and two hydroponic treatments (tapwater, T4; Arnon and Hoagland nutrient solution, T5). Although in chamber growth conditions the acclimatized plants showed statistical differences in several growth parameters depending on the treatments, no differences were observed in the survival percentage. Shoot and root fresh and dry weight and leaf area were highest in T5 and lowest in T2 and T4. Field survival differed significantly between treatments, discriminating a group with high survival rates (T5: 73.3%, T3: 86.7%, and control treatment: 100%) and another with low survival rates (T2: 33.3%; T1: 35% and T4: 36.7%). At harvest, there were no significant differences in the total fresh weight. However, the percentage of biomass partitioned to roots was significantly higher in T3 and T5, which resulted in a higher tuberous roots yield than that of the control treatment.

Keywords: acclimatization, cassava, in vitro plant regeneration, Manihot esculenta, tuberous roots, yields.

INTRODUCTION

Cassava (*Manihot esculenta* Crantz) is a staple crop with great economic importance, constituting a basic component in the dietary of over 1,000 million people in tropical and subtropical countries (FAO/FIDA, 2000; Ceballos, 2002).

The broad diversity of uses that can be given to the whole plant, its flexibility regarding the timing of planting and harvesting, and its ability to be produced under a wide range of edaphoclimatic conditions (Puonti-Kaerlas, 1998; Ceballos, 2002) make this species fulfill a prominent role in the context of food security (FAO/FIDA, 2000).

World production of the cassava tuberous root is estimated to be 214 million tons (FAO, 2009). In Argentina, cassava is cultivated mainly by small scale and subsistence farmers in the Northeast region, reaching a tuberous root production of 200,000 tons (De Bernardi, 2001).

Traditionally, cassava is propagated asexually by stem cuttings. Although this is a simple and inexpensive method that simultaneously preserves the varietal features, it provides very low multiplication rates and it is the main mode of transmission and spread of pests and diseases, thus affecting the quality and quantity of the planting material and crop yield (Roca and Jayasinghe, 1982; Puonti-Kaerlas, 1998; Pedroso de Oliveira et al., 2000; Bellotti et al., 2002; Albarrán et al., 2003). Therefore, to ensure regional development of this crop, it is necessary to multiply plants from breeding lines, elite cultivars or systemic pathogen-free materials (Smith et al., 1986; Puonti- Kaerlas, 1998; Thro et al., 1999).

In this regard, the techniques of *in vitro* tissue culture can provide solutions to these problems. The cassava meristem culture has allowed the propagation of virus and other systemic pathogens-free plants (Roca et al., 1991), while the uninodal segment culture has facilitated the rapid plant multiplication and *in vitro* conservation of different varieties (Smith et al., 1986; Roca et al., 1991; Pedroso de Oliveira et al., 2000; Albarrán et al., 2003). According to Roca et al. (1991), the use of this technique allows obtaining three to five plants per month from each nodal segment. It should be noted that the efficiency of the technique varies with the genotype, as some clones are more easily adapted to the process of *in vitro* propagation than others (Pedroso de Oliveira et al., 2000; Albarrán et al., 2003).

The most critical phase of micropropagation is acclimatization, which is the gradual transfer of plants from the *in vitro* environment to *ex vitro* conditions, where they undergo a lower relative humidity and a comparatively much higher light intensity (Pospíšilová et al., 1999; Jorge et al., 2000). Moreover, these plants have a lower nutrient availability, mechanical damage to the roots (Segovia et al., 2002) and exposure to saprophytic and eventually phytopathogenic microorganisms (Grattapaglia and Machado, 1990).

Several studies have reported a high mortality of cassava plants in the acclimatization stage (Zok et al., 1993; Da Silva et al., 1995; Azcón Aguilar et al., 1999; Jorge et al, 2000; Zimmerman et al ., 2007, Marín et al. 2008). Mortality can reach 95% if the right technology is not used (Segovia et al., 2002). According to Jorge et al. (2000), the low survival in the acclimatization phase and field establishment may be one of the reasons why *in vitro* tissue culture has not been adopted as a tool for the propagation of cassava in a large scale.

It is important to note that while production costs of plants increase with the implementation of these techniques, *in vitro* tissue culture has the potential to produce large numbers of quality plants at any time of the year, hundreds of times faster than traditional techniques (Roca, 1984; Thro et al., 1999; Ceballos, 2002).

The aim of this chapter was to evaluate the *in vitro* establishment and multiplication of 28 cassava clones of agronomic interest for the Northeast region of Argentina and to assess the effect of different acclimatization treatments on plant survival and growth of the EC118 clone in growth chamber and field conditions. In addition, we determined the tuberous root yield of acclimatized plants in comparison with the traditional planting technique.

MATERIALS AND METHODS

The plant material consisted of 28 clones of cassava (*Manihot esculenta* Crantz) provided by the germplasm bank of the "Instituto Nacional de Tecnología Agropecuaria, Estación Experimental Agropecuaria (INTA EEA)" El Colorado (Formosa, Argentina) (Table 1).

In Vitro Establishment and Multiplication

Cassava (*Manihot esculenta* Crantz) stem cuttings with approximately six buds were planted in pots with a mixture of 1:1 of black soil and fine sand to improve drainage, and then maintained under greenhouse conditions. Uninodal segments (approximately 10 mm long) dissected from cassava plants grown under greenhouse conditions were used as source of explants for all experiments. Explants were disinfected in 70% ethanol for 1 min and then immersed in 1.1% sodium hypochlorite plus 0.05% (v/v) Triton X-100® for 20 min, and finally rinsed three times with sterile distilled water.

Table 1. Origin and main features of cassava clones provided by the germplasm bank of the INTA EEA El Colorado (Formosa, Argentina)

Clone	Origin	kg/plant	Plant height (m)	Root color
EC 22	Col. Drifin Porá, Corrientes	2.20	2.85	white
EC 107	EEA EC Mariño	3.03	2,80	brown
EC 44	Picada Sur Javier, Misiones	3.30	2.45	white
EC 42	Paraje López, Misiones	5.33	2.25	brown
EC 6	Monte Caseros, Corrientes	2.93	3.00	white
EC 20	Ingeniero Juárez, Formosa	1.76	2.30	brown
EC 88	Cahuare, Misiones	3.70	2,00	brown
EC 19	Paso Ita, Corrientes	1.73	1.60	white
EC 26	Col. San Justo, Corrientes	1.66	1.50	brown
EC24-10	Col. Drifin Pora, Corrientes	1.92	1.50	brown
EC 29-9	Col. San Justo, Corrientes	2.03	1.70	brown
EC 3	Col. Santa Ana, Corrientes	3.16	2.10	brown
EC 90	Cahuare, Misiones	4.20	1.90	brown
EC 157	Manantiales, Corrientes	2.40	1.90	brown
EC 165	Manantiales, Corrientes	2.90	1.80	white
EC 161	Yrigoyen, Formosa	2.66	1.90	brown
EC 110	Pomberí	2,73	2.00	white
EC 111	Campeona C.B.	2.86	1.80	brown
EC 74	Km 1124 San Javier, Ctes	1.80	2.05	white
EC 113	EEA, E.C., Negra, Formosa	1.86	1.80	white
EC 23	Drifin Porá	0.90	1.30	white
EC 118	C.A.6 3, Misiones	2.16	1.80	white
EC 124	Yerutí	3.86	2.20	white
EC 1-1	Col. Santa Ana, Entre Ríos	1.86	3.00	white
EC 121	C.A. 25.1, Misiones	3.00	2.90	brown
EC 163	Manantiales, Corrientes	1.16	2.20	white
EC 27-4	San Justo, Corrientes	3.73	2.00	brown
EC 162	No characterization	-	-	-

Resource: EEA INTA El Colorado. Unpublished data.

Uninodal segments were cultured aseptically on Murashige and Skoog (1962) basal medium (MS) additioned with 0.01 mg/L naphthaleneacetic acid (NAA), 0.01 mg/L 6-benzylaminopurine (BAP) and 0.1 mg/L gibberellic acid (GA$_3$). The culture media pH were adjusted to 5.8 with KOH and/or HCl and solidified with 0.75% agar (Sigma® A1296). Test tubes (43 mL capacity) were covered with aluminium foil and autoclaved at 1.46 kg cm^{-2} for 20 min. The cultures were covered with Resinite AF-50® film (Casco S. A. C. Company, Buenos Aires) and incubated in a growth chamber at 27±2°C under a 14 h photoperiod regime with an irradiance of 116 $\mu mol.m^{-2}.s^{-1}$ provided by cool white fluorescent lamps.

After 30 days of *in vitro* establishment, we evaluated the percentage of contamination, sprouting and establishment, as well as plant height, number of nodes per plant, rooting percentage and average root number per plant. An explant was considered as established when it was alive and disinfected after 30 days of culture.

Subsequently, regenerated plants were subjected to a first multiplication cycle. *In vitro* uninodal segments were cultured in MS supplemented with 0.01 mg/L NAA, 0.01 mg/L BAP and 0.1 mg/L GA_3, and after 30 days the parameters evaluated in the establishment phase were assessed again.

Acclimatization of Plants in Growth Chamber

Fifty days after the second multiplication cycle, *in vitro* plants of the EC118 cultivar were removed from test tubes, soaked in tapwater to remove the remaining culture medium and rinsed carefully. Then, they were submerged in fungicide solution (2% w/v Captan®) for 10 min, and subjected to five treatments of acclimatization (Table 2): three consisting in the use of different solid substrates placed in 180 cm^3 plastic pots (Fig. 1A), and two conducted under hydroponic conditions, with a device designed to allow autonomous and constant aeration (Figs. 1 B and C).

Table 2. Description of the acclimatization treatments of cassava
***in vitro* plants (EC118 cultivar)**

Treatments		Conditions
Solid substrates	T_1	Perlite
	T_2	sand + 3% (w/w) vermicompost
	T_3	D1 Dynamics® a commercial substrate composed of peat and perlite, 9:1 respectively
Hydroponic system	T_4	Tapwater
	T_5	Arnon and Hoagland (1940) nutrient solution

In all treatments, plants were protected by transparent plastic covers (like a moist chamber) to avoid dehydration, and kept in a growth chamber at $27\pm2°C$ with 14 h photoperiod and an irradiance of 215 $\mu mol.m^{-2}.s^{-1}$ for 20 days. The relative humidity was gradually decreased until the complete elimination of the plastic cover. Plants of treatments 1, 2 and 3 were drenched with fungicide (2% w/v Captan®) and irrigated with 20 cm^3 of Arnon and Hoagland (1940) nutrient solution per plant, three times a week, while plants of the hydroponic treatments maintained the initial level nutrient solution or tapwater as appropriate.

The variables evaluated were: survival rate, plant height, number of nodes per plant, leaf area (measured with a portable meter L-3000C LI-COR®), and fresh and dry aerial and radical weight, total fresh and dry weight, and dry matter percentage for both *in vitro* regenerated plants (initial state) and acclimatized plants incubated for 20 days in growth chamber (final state).

Figure 1. Acclimatization phase in growth chamber: (A) Treatment of acclimatization in multicell trays filled with three different types of substrates and covered with a transparent plastic. (B) Treatment of acclimatization under hydroponic conditions. (C) Detail of the device used to promote aeration in the hydroponic system.

Survival and Growth Performance under Field Conditions

Field trials with acclimatized plants of the EC118 clone were carried out at the experimental farm of the Universidad Nacional del Nordeste (UNNE), Corrientes (27°28´S, 58°16´W), Argentina. The soil of the experimental site is a mixed hyperthermic alfic Udipsamment (Escobar et al., 1994). The climate of this zone is subtropical. This zone has an average annual precipitation of 1500 mm, which represents a positive hydric balance, an average annual temperature of 21.5°C with an average minimum temperature of the coldest month (July) between 13 to 16°C and a frost-free period of 320 to 360 days. The annual frost frequency is 0.4 (between June and July) and no frost probability expected between October and April (Bruniard, 1999).

Donor plants of the control stem cuttings were taken from a clonal orchard of the Facultad de Ciencias Agrarias (UNNE), where they were grown under the same environmental and agrotechnical conditions.

At this phase, the performance of the field acclimatized plants was compared to plants grown from stem cuttings with three-four nodes (control treatment; Tc). The plant spacing for all treatments was 1 m x 1 m. Plants were irrigated after planting and then three times a week during the first three weeks. All plants were harvested at 5 ½ months after planting.

Field survival was recorded at 20, 30, 60, 90, 120 and 165 days after planting. The following parameters were evaluated at harvest time: plant height, main branch length, number of branches per plant, number of nodes per plant, percentage of branched plants and shoot and tuberous root fresh weight per plant. The yield of each treatment was expressed as a percentage of the yield obtained in the control treatment (Tc).

Experimental Design and Statistical Analysis

The experimental design of the *in vitro* establishment and multiplication phase of the 28 clones of cassava, and the acclimatization of the EC118 clone in growth chamber was completely randomized, with three repetitions for each treatment. For the evaluation of plant performance and behavior under field conditions, the experimental design was a randomized complete block with three replications for each treatment. Data were subjected to one-way analysis of variance (ANOVA) after verifying the normality of the variables, and the means were compared by Duncan's multiple comparison test ($p \leq 0.05$;

p≤0.01), using InfoStat software professional version 1.1 (InfoStat, 2002). Means are presented with standard error (±SE).

RESULTS AND DISCUSSION

In Vitro Establishment

All cassava clones evaluated were established *in vitro*. The *in vitro* establishment percentage varied between 38.7% and 100%. Fungal and bacterial contamination was observed in 89% of the clones. Although fungal infection was more frequent, affecting 71% of the clones, it ranged between 4.6% and 18%, while bacterial infection reached 56.4%, affecting fewer clones (51%). The percentage of unresponsive explants ranged between 0% and 6%, with the exception of the EC 19 clone, which presented 20% of unresponsive explants.

In the establishment phase, the percentage of sprouting showed significant differences between clones (p≤0.05), ranging between 38.7% and 95.5%. Almost 90% of the clones tested exceeded 50% of sprouted explants, while more than half of the clones (61%) had values above 75% (Fig. 2).

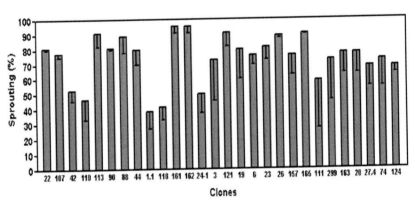

Figure 2. Sprouting percentage of the 28 cassava clones after 30 days of culture under *in vitro* conditions.

We found highly significant differences between clones with respect to plant height (p≤0.01), ranging between 3.3 and 69.4 mm; clones EC29 and EC26-9 presented the shortest plants, whereas EC121, EC107 and EC20 the tallest ones. The number of nodes per plant also differed significantly between clones (p≤0.01), with average values between 2.4 and 9.7 nodes. In general, clones that produced the shortest plants were those that formed the lowest number of nodes per plant, and clones that produced the tallest plants were those that formed the largest number of nodes per plant.

Rooting was observed in 93% of the clones during the establishment phase. In addition, significant differences in the rooting percentage, which ranged between 6.3% and 91%, were found between clones (p≤0.01).

It should be noted that 57% of the clones showed a rooting percentage above 50%. Only two clones (EC26 and EC299) were not rooted in the establishment phase. The average root number per plant (p ≤0.01) also differed between clones, ranging between 0 and 4 roots per plant.

In Vitro **Multiplication**

In the multiplication phase, 100% of the *in vitro* established clones regenerated plants from the uninodal segment culture. Maximal sprouting (100%) was found in 71% of the clones, while the remaining clones recorded values above 77% (Fig. 3). These values are higher than those reported by Albarrán et al. (2003), who obtained plants in the 34 clones evaluated, reporting percentages of regeneration between 60 and 90%. Acedo (2002) obtained 100% plant regeneration of the cassava cultivar "Golden Yellow" when using MS medium free of plant growth regulators (PGRs), and low levels of regeneration (13-21%) when using MS supplemented with NAA, BAP and GA$_3$.

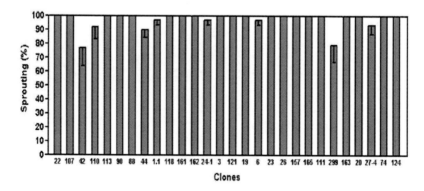

Figure 3. Sprouting percentage of the 28 cassava clones after 30 days of the first *in vitro* multiplication.

In our study, the plants were morphologically normal in phenotype (size, shape and color) independently of the clones tested. Therefore, we can state that the plants established *in vitro* are phenotypically stable compared to the original material.

Plant height varied significantly between clones (p ≤0.01), with average values between 2.9 mm and 57 mm, slightly shorter than those observed in the establishment phase. In general, the clones that had the tallest stem in the establishment phase also developed the tallest stem in the multiplication phase. The number of nodes per plant also differed significantly between clones (p≤0.01), with average values between 2.3 and 6.6 nodes per stem. Marín et al. (2008) reported plant height values of 30 to 70 mm and a node number per plant of 3.5 and 7, for 19 elite cassava clones after 60 days of culture (twice the time used in our experiment) in a culture medium consisting of ⅓ MS and 0.02 mg/L NAA. Smith et al. (1986) reported a significant increase in the average node number per plant in relation to the addition of PGRs in the basal medium MS, being the addition of NAA and BAP more effective than IBA alone. These authors explained this increase in the number of nodes per plant as a consequence of the promotion of multiple shoots. On the other hand, Pedroso de Oliveira et al. (2000) reported an increase in plant height in the first cycle of multiplication from 8.6 to 17.7 mm for six varieties of cassava from Brazil, using a culture medium composed of 35% macronutrients and micronutrients of MS supplemented with 1 mg/L thiamine, 100 mg/L inositol, 0.01 mg/L NAA, 0.01 mg/L GA_3 and 2 % sucrose. According to these authors, the presence of poorly developed plants was more frequent than the presence of etiolated plants.

In our experiment, all the clones tested were able to root in the multiplication phase, the rooting percentage ranging between 20% and 100%. It should be noted that 89% of the rooting clones had values above 50% (Fig. 4). The average root number per plant also differed significantly between clones (p≤0.01), ranging from 1 to 5 roots. Pedroso de Oliveira et al. (2000) reported rooting in the first cycle of multiplication in four of the six clones tested, with values between 30 and 70%, reaching 100% effectiveness just in the third cycle of multiplication.

In the multiplication phase, clones that showed higher plant height were those reporting the highest average root number per plant like in the establishment phase. Pedroso de Oliveira et al., (2000) argued that the presence of roots in cassava seedlings is beneficial to the multiplication process because it promotes the absorption nutrients and therefore a good production of buds that will serve as explants for the following culture cycles.

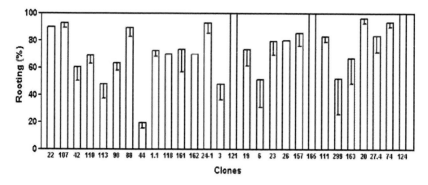

Figure 4. Percentage of rooted explants of 28 cassava clones grown *in vitro* 30 days after the first multiplication.

In general, the results indicated a great variability between clones (Fig. 5) with respect to all the parameters evaluated. This pronounced effect of the genotype on *in vitro* cassava plant development has been reported by other authors (Roca, 1984; Pedroso de Oliveira et al., 2000; Albarrán, et al., 2003; Marín et al., 2008). In this regard, Smith et al. (1986) proposed adjustments to the culture medium to stimulate the growth of cassava varieties that have low multiplication efficiency.

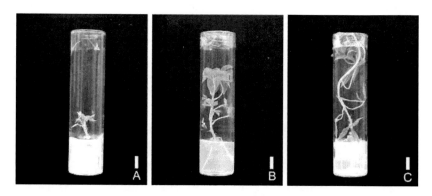

Figure 5. Variability of cassava clones with respect to plant height, node and root number per plant after 30 days of *in vitro* establishment. (A) clone EC24 (B) clone EC161, (C) clone EC121. Bar: 10 mm.

Acclimatization of Plants in Growth Chamber

The *ex vitro* survival, recovery and growth of cassava plants obtained *in vitro* were successful, regardless of the acclimatization method assayed. The survival rate of plants at the end of the acclimatization phase in growth chamber varied between 96% (T_1, T_4 and T_5) and 100% (T_2 and T_3). These values were similar to those reported by Pedroso de Oliveira et al. (2000) for six Brazilian cassava varieties (92%) and higher than those obtained by Broomes and Lacon (1995) at the end of the first week of acclimatization of plants regenerated *in vitro* in liquid medium (82%). The values of our results are also higher than those found by Azcón Aguilar et al. (1997), who inoculated cassava plants with *Glomus desserticola* as a strategy to increase *ex vitro* survival during the acclimatization phase (75%). Zimmerman et al. (2007) acclimatized cassava plants rooted in vermiculite instead of gelling agent, and reported survival values greater than 95%; probably, this way to promote rhizogenesis provides the possibility to preserve the integrity of roots when removing them from the substrate.

Zok et al. (1993) analyzed the survival of acclimatized cassava plants using different combinations of soil with vermiculite, sawdust and coffee husks, using soil alone as control treatment. These authors reported survival values of 5% to 55% and obtained better performance with the mixture of soil with vermiculite, and remarked the importance of providing the substrate with good aeration and water retention so as to achieve successful plant acclimatization.

Le et al. (2007) achieved 93% survival after maintaining the plants for 7 days in water and then 20 to 25 days in a nutrient solution. Using a similar procedure but then transplanting plants to pots, Marín et al. (2008) obtained only between 0% and 57% survival for 19 elite cassava clones. However, Albarrán et al. (2003) reported survival values above 50% in 32 of the 34 clones of cassava plants evaluated following the method used by Marín et al. (2008). In our study, we observed significant differences between acclimatization treatments ($p \leq 0.01$) with respect to all the parameters evaluated. Higher values were obtained in plants growing under hydroponic conditions with Arnon and Hoagland nutrient solution (T_5) (Table 3).

All parameters, except for radical fresh weight, showed significant differences among the initial (T0) and final condition after the different treatments of acclimatization (Table 3). Radical fresh weight differed significantly ($p \leq 0.01$) from the initial condition (T0) only when plants were grown under hydroponic conditions with Arnon and Hoagland nutrient solution (T_5). The number of nodes per plant and leaf area was high in T_5 and T_3, differing significantly from T0, T_2 and T_4 (Table 3). These data are in contrast to those reported by Da Silva et al. (1995), who found no significant differences in survival, leaf area and node number, using different substrates for the acclimatization of cassava plants. These authors obtained survival values ranging between 49% and 56% and inferred a relationship between survival and age of the rooted shoots at the beginning of the acclimatization process.

Table 3. Effects of different acclimatization treatments of cassava plants (clone EC118) on several growth parameters, evaluated at 20 days of *ex vitro* culture, with respect to initial *in vitro* conditions (T0): plant height (PH), number of nodes per plant (NNP), leaf area (LA), aerial fresh weight (AFW), radical fresh weight (RFW), total fresh weight (TFW), aerial dry weight (ADW), radical dry weight (RDW), total dry weight (TDW) and dry matter (DM)

Treatment	PH (cm)	NNP	LA (cm^2)	AFW (g)	RFW (g)	TFW (g)	ADW (g)	RDW (g)	TDW (g)	DM (%)
T_0	8.41[a]	7.67[a]	7.08[a]	0.18[a]	0.12[a]	0.30[a]	0.02[a]	0.01[a]	0.04[a]	11.75[a]
T_1	11.31[b]	11.56[cd]	33.57[cd]	0.63[bc]	0.19[a]	0.82[b]	0.09[b]	0.03[b]	0.13[b]	14.12[ab]
T_2	10.38[b]	10.78[bc]	17.86[b]	0.42[b]	0.15[a]	0.58[b]	0.07[b]	0.03[b]	0.10[b]	18.02[c]
T_3	11.89[b]	12.11[de]	44.26[d]	0.77[c]	0.17[a]	0.94[b]	0.11[b]	0.02[b]	0.13[b]	13.62[ab]
T_4	11.02[b]	10.00[b]	21.12[bc]	0.50[bc]	0.16[a]	0.66[b]	0.10[b]	0.02[b]	0.12[b]	19.20[c]
T_5	14.44[c]	13.22[e]	90.66[de]	1.93[d]	0.44[b]	2.37[c]	0.30[c]	0.07[c]	0.36[c]	15.12[b]

Different letters within columns indicate significant differences ($p \leq 0.01$).

Figure 6 shows the appearance of *in vitro* plants (T0) which were subjected to different acclimatization procedures (T_1 to T_5). In addition, it is possible to observe that the commercial substrate Dynamics® (T_3) and the hydroponic treatment with nutrient solution of Arnon and Hoagland (1940) (T_5) were beneficial on the vegetative growth, and that the hydroponic treatment improved radical development remarkably.

The use of the commercial substrate Dynamics® (T_3) and nutrient solution (T_5) led to an increase of 3.3 and 9 times in the *in vitro* dry weight values, respectively (Fig. 7). This significant increase in the biomass of acclimatized cassava plants is consistent with the findings of Pospíšilová et al. (1999), who worked with *Nicotiana tabacum*. These authors argued that if the *ex vitro* transplantation is successful, it ensures higher plant growth.

The percentage of dry matter showed significantly higher values ($p \leq 0.01$) in plants derived from sand + vermicompost (T_2) and hydroponic conditions using tapwater (T_4). However, both treatments had lower values of total dry weight (Fig. 7), implying a lower water content in plants subjected to these treatments. This decrease in dry weight accompanied by an increase in the percentage of dry matter was informed by Clostre and Suni (2007) in *Lemna gibba* L. and by Gerardeaux et al., (2009) in *Gossypium hirsutum* L., in both cases associated with a lower potassium content in the growth medium.

Figure 6. Cassava plants (clone EC118) at the beginning and at the end of the acclimatization phase in growth chamber. *In vitro* plant (initial state or T0), acclimatized plant using solid substrates as perlite (T_1), sandy ground + vermicompost (T_2), commercial mixture of peat and perlite (Dynamics ®) (T_3), and acclimatized plant using hydroponic treatment with tapwater (T_4), and Arnon and Hoagland nutrient solution (T_5).

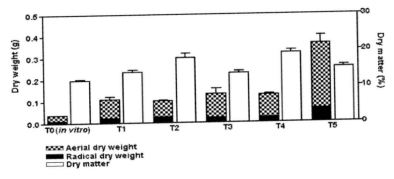

Figure 7. Aerial and radical dry weight and percentage of dry matter of the cassava plants derived from different treatments at the end of the acclimatization phase (clone EC118) in growth chamber, with respect to the initial condition (T0).

Survival and Growth Performance under Field Conditions

Field survival of plants showed significant differences between treatments at all the dates evaluated ($p \leq 0.05$), being higher in plants derived from stem cutting (Tc) and acclimatized plants in commercial substrate (T_3) and hydroponically with nutrient solution (T_5).

The largest reduction in the percentage of plant survival was remarked between 20 and 30 days of the transplantation field, and then remained constant up to 165 days of culture (Fig. 8). For this reason, we consider that the values recorded at 30 days of planting would be most appropriate to compare the survival of cassava plants acclimatized with different treatments. Two groups were distinguished: one with high survival rates (73.3%, 86.7% and 100% for T_5, T_3 and Tc, respectively) and another group with low survival rates (T_2: 33.3%; T_1: 35% and T_4: 36.7%).

Albarrán et al. (2003) reported values of field survival of acclimatized cassava plants between 22% and 100% depending on the clones, and emphasized that plants that survive the first two months in field conditions have a high probability to achieve productive age. This reduction in the survival rate was also observed in seed-derived cassava plants that were transplanted under field conditions, because for two to three months of culture they are more fragile than stem cutting-derived plants (Ceballos et al., 2002).

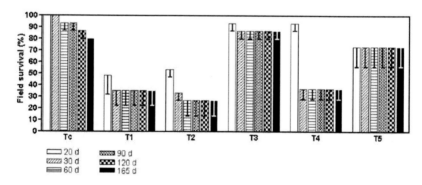

Figure 8. Field survival of cassava plants (clone EC118) acclimatized with different treatments compared to the control treatment (plants derived from stem cutting) during the crop cycle.

There were no significant differences between treatments ($p \leq 0.05$) with respect to the number of nodes per plant, main branch length, number of branches per plant and percentage of branched plants.

Fresh weight of the tuberous root, stem and leaf fresh weight and total fresh weight as well as the partition between tuberous roots and aerial organs showed significant differences between treatments (p≤0.05). Although plants grown from stem cuttings yielded the highest values of total fresh weight, they showed the lowest tuberous root fresh weight.

Table 4. Aerial fresh weight (AFW), radical fresh weight (RFW) and total fresh weight (TFW) of cassava plants (clone EC118) acclimatized with different treatments, transplanted under field conditions and evaluated at 165 days after planting Tc (plants derived from stem cutting used as control treatment)

Treatment	AFW (g)	RFW (g)	TFW (g)	Partitioning to stem + leaves (%)	Partitioning to tuberous roots (%)
T_c	702.92^c	188.33^a	891.25^b	79.23^c	20.77^a
T_1	417.17^{ab}	135.38^a	552.54^{ab}	67.77^b	22.33^a
T_2	287.50^{ab}	130.00^a	417.50^a	77.67^c	32.23^b
T_3	399.17^{ab}	295.42^b	694.58^{ab}	57.09^a	42.91^c
T_4	268.75^a	112.50^a	381.25^a	68.32^b	31.68^b
T_5	531.25^{bc}	303.33^b	834.58^b	62.33^{ab}	37.67^{bc}

Different letters within columns indicate significant differences (p≤0.05).

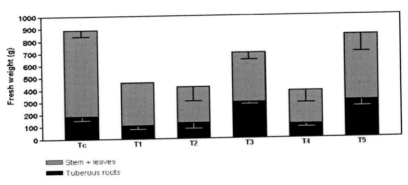

Figure 9. Aerial and radical fresh weight of cassava plants (clone EC118) acclimatized with different treatments and transplanted under field conditions, compared to control treatment (Tc: plants derived from stem cuttings).

On the other hand, total fresh weight of the acclimatized plants under treatments T_3 and T_5 did not differ significantly with respect to the control treatment (Tc), whereas tuberous root fresh weight showed statistical differences in both absolute value and percentage weight partitioned to tuberous root (Table 4, Fig. 9). Probably as a consequence of late transplanting of plants (November) and a shortening of the crop cycle (5.5 months), it was not possible to achieve the tuberous root yield expected for the EC118 clone (Table 1).

Significant differences in tuberous root yield (expressed as a percentage of yield obtained in the control treatment) were observed between treatments ($p \leq 0.05$). The highest yield values were achieved in plants acclimatized in commercial substrate (T_3) and in hydroponic nutrient solution (T_5), which were 57% and 61% above the yield achieved by plants derived from stem cutting used as control treatment, respectively (Fig. 10).

Cassava plants acclimatized with T_3 and T_5 showed a behavior similar to those derived from stem cuttings. It is likely that the highest yield observed in these treatments was due to the higher initial development of these plants compared to other treatments and the slow growth typically shown by plants derived from stem cuttings under field conditions. According to Alves (2002), true leaves begin to expand just 30 days after stem cutting planting, at which photosynthesis begins to contribute positively to plant growth. For this reason, both tuberization and photoassimilate translocation would start later than in acclimatized plants, thus determining their production.

While plant survival in the acclimatization phase in growth chamber did not differ between treatments, the differences observed in total weight, leaf area and root biomass at the end of this stage confirm that the plants obtained were significantly different, a condition that resulted in differential responses in the field plant survival.

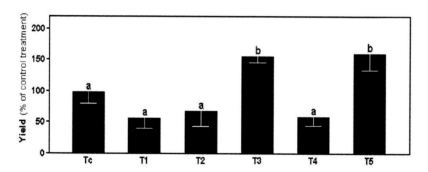

Figure 10. Tuberous root yield of cassava plants (clone EC118) derived from different acclimatization treatments compared to control treatment (Tc: plants derived from stem cuttings). The yield is expressed as a percentage of yield obtained in the control treatment).

It is likely that the greatest root development obtained with T_3 and T_5 favors further exploration of the soil, allowing better absorption of water and nutrients in field conditions compared to other treatments, including the control one.

CONCLUSION

- The basal medium MS supplemented with 0.01 mg/l NAA, 0.01 mg/l BAP and 0.1 mg/l GA_3 allowed the *in vitro* establishment, regeneration and multiplication of 28 clones from uninodal segment culture with optimal growth.

- In the establishment phase, *in vitro* rooting was achieved in 93% of the clones, of which 57% had a rooting percentage above 50%. In the first cycle of multiplication, all tested clones rooted and 89% of them showed a rooting percentage above 50%.

- It was possible to observe a pronounced effect of genotype on *in vitro* development of cassava plants. The average node number per plant at 30 days of culture ranged between 2.3 and 6.6 for the different clones during the first cycle of multiplication.

- Clones EC121, EC107 and EC20 were those with the best behavior *in vitro* both in the establishment and multiplication phases considering all the parameters evaluated.

- Acclimatization in growth chamber was successful, and a high survival of plants was found with all treatments (96% to 100%). Acclimatization treatments caused significant differences in leaf area and dry weight of aerial and radical parts, which resulted in a differential development of plants that subsequently affected the response of field grown plants, resulting in a higher yield of tuberous roots in plants acclimatized with the commercial substrate and with the nutrient solution in hydroponic conditions than that of plants derived from stem cuttings.

- With regard to field survival, it was possible to distinguish two groups: a group with high field survival (plants derived from stem cuttings and from acclimatization treatments with commercial substrate and hydroponic nutrient solution) and another group with lower field survival (plants derived from the treatments with sand + vermicompost, perlite and hydroponic tapwater).

- Although there were no differences with respect to the total fresh weight, the percentage of biomass partitioned to tuberous roots was significantly higher in the treatments with commercial substrate (T₃) and in hydroponic nutrient solution (T₅), which resulted in a higher yield tuberous root compared to plants derived from stem cuttings.
- The *in vitro* propagation process from the dissection of uninodal segments to field planting is carried out in approximately 10 weeks. Therefore, *in vitro* propagation is a promising alternative for the multiplication of cassava plants in Argentina, because it allows propagation rates and field yields higher than those obtained by traditional propagation methods.

ACKNOWLEDGMENTS

The authors are grateful to CONICET, Secretaría General de Ciencia y Técnica (UNNE) for the financial support and INTA EEA El Colorado for providing the plant material. The authors wish to express their gratitude to María Victoria González Eusevi, for her valuable comments on the manuscript.

REFERENCES

Acedo, V. Z. 2002. Meristem culture and micropropagation of cassava. *Journal of Root Crops* 28: 1-7.

Albarrán, J.; F. Fuenmayor and M. Fuchs. 2003. Propagación clonal rápida de variedades comerciales de yuca mediante técnicas biotecnológicas. Revista Digital del Centro Nacional de Investigaciones Agropecuarias de Venezuela. CENIAP Hoy N°3. www.ceniap.gov.ve/ceniaphoy/articulos/n3/texto/ albaran.htm [Fecha de consulta: 11/11/2010].

Alves, A. A. C. 2002. Cassava Botany and Physiology. In: R. J. Hillocks; J. M. Thresh and A. C. Bellotti (eds.), Cassava Biology, Production and Utilization. *CABI Publishing*, New York, USA, p 67-89.

Arnon, D. I. and D. R. Hoagland. 1940. Crop production in artificial culture solutions and in soils with special reference to factors influencing yields and absorption of inorganic nutrients. *Soil Science* 50: 463-83.

Azcón Aguilar, C.; M. Cantos; A. Troncoso and J. M. Barea. 1997. Beneficial effect of arbuscular mycorrhizas on acclimatization of micropropagated cassava plantlets. *Scientia Horticulturae* 72: 63-71.

Bellotti, A.; W. Roca; J. Tohme; P. Chavarriaga; R. H. Escobar and C. J. Herrera. 2002. Biotecnología para el manejo de plagas en la producción de semilla limpia. In: B. Ospina and H. Ceballos (eds.), La yuca en el tercer milenio: sistemas modernos de producción, procesamiento, utilización y comercialización. *CIAT*, Cali, Colombia, p. 255-261.

Bromees, V. F. and R. Lacon. 1995. Influence of medium components on hardening of cassava after micropropagation in liquid nutrient medium. In: Proceedings of the Second International Scientific Meeting of the Cassava Biotechnology Network, Bogor, Indonesia, p. 210-219.

Bruniard, E. 1999. Los regímenes hídricos de las formaciones vegetales. Aporte para un Modelo Fitoclimático Mundial. *EUDENE*, Resistencia, Argentina. 382 p.

Ceballos, H. 2002. La Yuca en Colombia y el mundo: Nuevas perspectivas para un cultivo milenario. In: B. Ospina and H. Ceballos (eds.), La yuca en el tercer milenio: sistemas modernos de producción, procesamiento, utilización y comercialización. *CIAT*, Cali, Colombia, p. 1-13.

Clostre, G. and M. Suni. 2007. Efecto del nitrógeno, fósforo y potasio del medio de cultivo en el rendimiento y valor nutritivo de *Lemna gibba* L. (Lemnaceae). *Revista Peruana de Biología* 13: 231-235.

Da Silva, A. T.; M. Pasqual; J. S. Ishida and L. E. C. Antunes. 1995. Aclimatação de plantas provenientes da cultura *in vitro*. *Pesquisa Agropecuária Brasileira* 30: 49-53.

De Bernardi, L. A. 2001. Cadenas Alimentarias: Fécula de mandioca. *Revista Alimentos Argentinos* N° 17.

Escobar, E. H., O. Ligier; R. Melgar; M. Matteio and O. Vallejos. 1994. Mapa de suelos de los Departamentos de Capital, San Cosme e Itatí de la Provincia de Corrientes. *INTA/CFI/ICA*, 125 p.

FAO. 2009. FAOSTAT. Food and Agriculture Organization. Roma, Italia. http://faostat.fao.org/site/567/default.aspx [Fecha de consulta: 11/11/10].

FAO/FIDA. 2000. La economía mundial de la yuca: hechos, tendencias y perspectivas. *Food and Agriculture Organization / Fondo Internacional de Desarrollo Agrícola*, Roma, Italia, 59 p.

Gerardeaux, E.; E. Saur; J. Constantin; A. Porté and L. Jordan-Meille. 2009. Effect of carbon assimilation on dry weight production and partitioning during vegetative growth. *Plant Soil* 324: 329-343.

Grattapaglia, D. and L. Machado. 1990. Micropropagação. In: A. L. Torres and L. S. Caldas (eds.) Técnicas e Aplicações da cultura de tecidos de plantas. *ABCTP/Embrapa*, Brasil, p. 99-170.

InfoStat 2002. Infostat versión 1.1. Grupo Infostat, Facultad de Ciencias Agropecuarias, Universidad Nacional de Córdoba, Argentina.

Jorge, M. A.; A. I. Robertson; A. B. Mashingaidze and E. Keogh. 2000. How *in vitro* light affects growth and survival of *ex vitro* cassava. *Annals of Applied Biology* 137: 311-319.

Le, B. V.; B. L. Anh; K. Soytong; N. D. Danh and L. T. Anh Hong. 2007. Plant regeneration of cassava (*Manihot esculenta* Crantz) plants. *Journal of Agricultural Technology* 3: 121-127.

Marín, A.; D. Perdomo; J. G. Albarrán; F. Fuenmayor and C. Zambrano. 2008. Evaluación agronómica, morfológica y bioquímica de clones élites de yuca a partir de vitroplantas. *Interciencia* 33: 365-371.

Murashige, T. and F. Skoog. 1962. A revised medium for rapid growth and bioassays with tobacco tissue cultures. *Physiologia Plantarum* 15: 473-497.

Pedroso de Oliveira, R.; T. Da Silva Gomes and A. Duarte Vilarinhos. 2000. Avaliação de um sistema de micropropagação massal de variedades de mandioca. *Pesquisa Agropecuária Brasileira* 35: 2329-2334.

Pospíšilová, J; I. Tichá; P. Kadleček; D. Haisel and Š. Plzáková, 1999. Acclimatization of micropropagated plants to *ex vitro* conditions. *Biologia Plantarum* 42: 481-497.

Puonti-Kaerlas, J. 1998. Cassava Biotechnology. *Biotechnology and Genetic Engineering Reviews* 15: 329-364.

Roca, W. M. 1984. Cassava. In: W. R. Sharp; D. A. Evans; P. V. Ammirato and Y. Yamada (eds.), Handbook of Plant Cell Culture. Vol 2: Crop Species. *MacMillan Publishing*, Nueva York, p. 269-301.

Roca W. and U. Jayasinghe. 1982. El cultivo de meristemas para el saneamiento de clones de yuca. Guía de estudio. *CIAT*, Serie 04SC-02.05, Cali, Colombia. 47 p.

Roca, W. M.; B. Nolt; G. Mafla; J. Roa and R. Reyes. 1991. Eliminación de virus y propagación de clones en la yuca (*Manihot esculenta* Crantz). In: W. M. Roca and L. A. Mroginski (eds.), Cultivo de tejidos en la agricultura: fundamentos y aplicaciones. *CIAT, Cali*, Colombia, p. 403-420.

Segovia R. J.; A. Bedoya; W. Triviño; H. Ceballos; G. Gálvez and B. Ospina. 2002. Metodología para el endurecimiento masivo de vitroplantas de yuca. In: B. Ospina and H. Ceballos (eds.), La yuca en el tercer milenio:

sistemas modernos de producción, procesamiento, utilización y comercialización. *CIAT*, Cali, Colombia, p. 572-583.

Smith M. K.; B. J. Biggs and K. J. Scott. 1986. *In vitro* propagation of cassava (*Manihot esculenta* Crantz). *Plant Cell, Tissue and Organ Culture* 6: 221-228.

Thro, A. M.; W. M. Roca; J. Restrepo; H. Caballero; S. Poats; R. Escobar; G. Mafla and C. Hernández. 1999. Can *in vitro* biology have farmer-level impact for small-scale cassava farmers in Latin America? *In vitro Cellular and Developmental Biology Plant* 35: 382-387.

Zimmerman, T. W.; K. Williams; L. Joseph; J. Wiltshire and J. A. Kowalski. 2007. Rooting and acclimatization of cassava (*Manihot esculenta*) *ex vitro*. *Acta Horticulturae* 738: 735-740.

Zok, S.; L. M. Nyochembeng; J. Tambong and J. G. Wutoh. 1993. Rapid seedstock multiplication of improved clones of cassava (*Manihot esculenta* Crantz) through shoot tip culture in Cameroon. In: W. M. Rocca and A. M. Thro (eds.) Proceedings of the First International Scientific Meeting of the Cassava Biotechnology Network. Cartagena, Colombia, p. 96-104.

In: Cassava: Farming, Uses, and Economic Impact   ISBN:978-1-61209-655-1
Editor: Colleen M. Pace   © 2012 Nova Science Publishers, Inc.

Chapter 4

BIOTECHNOLOGICAL POTENTIAL OF CASSAVA RESIDUES: PEEL, BAGASSE AND WASTEWATER

Siddhartha G. V. A. O. Costa[a1], Marcia Nitschke[b] and Jonas Contiero[a]

[a]Department of Biochemistry and Microbiology, Biological Sciences Institute, UNESP Univ Estadual Paulista, Bela Vista, 13506-900, Rio Claro, SP, Brazil
[b]Department of Physical-Chemistry, Institute of Chemistry, University of São Paulo - USP, Av. Trabalhador São Carlense, 400, 13560-970, São Carlos, SP, Brazil

ABSTRACT

Advances in industrial biotechnology offer potential opportunities for the economic utilization of agro-industrial residues, such as those from the cassava processing industry. Three main types of residue are generated during the industrial processing of cassava: peels and bagasse (solid); and wastewater (liquid). Both types of waste are poor in protein content, but are carbohydrate-rich residues and generated in large amounts during the production of flour (which generates more solid

---

[1] Corresponding author. Address: UNESP Univ Estadual Paulista, Av. 24-A, 1515, Bela Vista, 13506-900, Rio Claro, SP, Brazil. Tel.: +55 19 35264180; fax: +55 19 35264176. E-mail address: sidd_georges@hotmail.com (S.G.V.A.O. Costa).

residues) and starch (which generates more liquid residues) from the tubers. Waste treatment and disposal costs constitute a huge financial burden to the cassava processing industry as well as an environmental problem. Therefore, there is a great need for the better management of these waste products. Due to its rich organic nature, cassava residue can serve as an ideal substrate for microbial processes in the production of different products. Attempts have been made to produce products such as organic acids, flavor and aroma compounds, mushrooms, methane and hydrogen gas, enzymes, ethanol, lactic acid, biosurfactants, polyhydroxy alkanoate, essential oils, xanthan gum and fertilizer from cassava bagasse, peels and wastewater. The use of cassava residues as feedstock in bio technological processes is a viable alternative that can contribute toward a reduction in production costs, an increase in the economic value of these residues and the minimization of environmental problems related to waste discharge. This study reviews processes and products developed for aggregating value to cassava residues through biotechnological means, demonstrating the potential of this agro-industrial raw material.

1. INTRODUCTION

In recent years, the spread of the application of agro-industrial residues has been considered an important strategy for facilitating the industrial development of microbial processes. A number of process have been developed using raw materials for the production of products such as organic acids, flavor and aroma compounds, methane and hydrogen gas, enzymes, ethanol, lactic acid and biosurfactants.

The choice of inexpensive agro-industrial residues is important to the overall economics of the process, leading to a reduction in the cost of the final product as well as an improvement in the economics of waste treatment. Waste disposal is a growing problem and new alternatives for the use of waste products should be studied. Biotechnology, especially technology involving fermentation and enzymes, has provided innovations for the use and development of new processes for agricultural feed stocks.

Three types of residue are generated during the industrial processing of cassava: peels and bagasse (solid); and wastewater (liquid). Both waste products are poor in protein content, but are carbohydrate-rich residues and generated in large amounts during the production of flour (which generates more solid residues) and starch (which generates more liquid residues) from the tubers. The use of cassava residues in biotechnological process has generated promising results. This paper discusses the processes, products developed and aggregated value of cassava residues through biotechnological means and future perspectives regarding the use of this important agro-industrial raw material.

2. CASSAVA

Cassava (*Manihot esculenta* Cranz) is the third most important source of carbohydrates in the tropics, following rice and maize. It is an important source of dietary calories for a large population in tropical countries of Asia, Africa and Latin America. World production of cassava has steadily increased from about 75 million tons in 1961-1965 to 153 million tons in 1991. Current estimated production is 160 million tons per year worldwide [1].

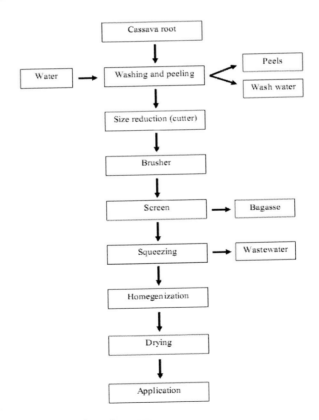

Figure 1. Industrial processing of cassava.

Industrial processing of cassava is performed mainly to isolate flour and starch from the tubers and most of factories are small or medium size. Figure 1 displays the processing of cassava tubers for the isolation of starch and flour. Most forms of cassava processing produce large amounts of waste, the type and composition of which are governed by the processing method and sophistication of the technology used. Cassava processing is generally considered to contribute significantly to environmental pollution and the depletion of water resources due to the strong, unpleasant odor and visual display of waste products. Some forms, especially starch extraction, also require large volumes of water [2].

2.1. Cassava Waste

Three types of residues are generated during the industrial processing of cassava: peels and bagasse (solid); and wastewater (liquid). Both waste products are poor in protein content, but are carbohydrate-rich residues and generated in large amounts during the production of flour (which generates more solid residues) and starch (which generates more liquid residues) from the tubers.

Solid waste is created by all forms of cassava processing. Inappropriate storage of solid waste for long periods is the main issue, especially with heavy rainfall culminating in the production of leachate, which can contaminate groundwater. Solid waste from cassava starch processing is divided into two categories: peels and bagasse.

2.1.1. Peels

Cassava peels represent 8 to 10% of the root dry matter. The peels contain a high amount of soluble carbohydrates (62%), low amounts of fiber (16%) and ash (4 to 6%) and a moderate level of nitrogen (1%). The use of cassava peels remains limited due to its low degree of digestibility as well as its toxicity stemming from high levels of hydrocyanic acid, thereby constituting a nuisance to the environment [3].

2.1.2. Bagasse

The processing of 250 to 300 tons of fresh cassava tubers results in about 280 tons of wet cassava bagasse. Cassava bagasse is made up fibrous root material and contains starch that could not be extracted through physical processing. Poor processing conditions may result in an even higher concentration of starch in cassava bagasse. Cassava tuber contains cyanide, but cassava bagasse does not have any cyanide content. It is poor in nitrogen content (2.3%), making it unattractive as an animal feed, and rich in carbohydrates (60%). Other components, such as calcium, potassium, lipids and phosphorous, are also found at low concentrations (< 1%). Cassava bagasse has a low ash content (1.44%), which offers numerous advantages in comparison to other crop residues, such as rice and wheat straw, for use in bioconversion processes involving microbial cultures [4].

2.1.3. Wastewater

Cassava wastewater comes from the pressing of cassava roots and is generated in large amounts during the production of cassava flour (1 ton of cassava flour generates 300 L of cassava wastewater). Due to its considerable organic content and the presence of cyanide, this wastewater is considered a pollutant. From another standpoint, however, it can serve as raw material for fermentative processes. Cassava wastewater is also considered a toxic waste for the presence of cyanogenic glucosides as linamarin and lotaustralin, naturally found in cassava. During the pressing operation almost all cyanogenic glucosides are carried to wastewater and after an acid or enzymatic hydrolysis of such glucosides cyanide and cyanidric acid are generated [5].

Table 1. Physiochemical composition of cassava wastewater [6]

Components	Concentration
Protein (%)	0.9
Fructose (g \l)	24.5
Glucose (g \l)	30.1
Maltose (g \l)	1.8
Nitrate (g \l)	0.7
Nitrite (mg \l)	0.05
Phosphorus (g \l)	0.9
Potassium (g \l)	3.9
Magnesium (g \l)	0.5
Sodium (mg\l)	23.1
Iron (mg\l)	6.1
Zinc (mg\l)	11.1
Manganese (mg\l)	4.1
Copper (mg\l)	14.1

The high sugar content of wastewater makes it suitable for bio-technological applications as a rich substrate for microbial growth and conversion. Concerning the cyanide presence, it is totally removed from wastewater (as well flour and starch) after heat treatment [5], i.e., an imperative operation to any fermentation process. Table 1 displays the physio-chemical composition of cassava wastewater.

The treatment and disposal costs of cassava residues constitute a huge financial burden to the cassava industry as well as an environmental problem. Therefore, there is considerable need for the better management of these residues. Use as feedstock in biotechnological processes is a viable alternative that can contribute toward a reduction in production costs as well as an increase in the economic value of these residues.

3. BIOTRANSFORMATION (PROCESS AND PRODUCTS)

In recent years, there has been an increasing trend toward the more efficient use of agro-industrial by-products for conversion into a range of value-added bio-products, including biofuels, biochemicals and biomaterials. Biotransformation based on the microbial conversion of agro-industrial residues can help solve environmental problems associated with their disposal as well as aggregate value to these residues and the products derived from their use. Due to its rich organic nature and low ash content, combined with the ease of its hydrolysis, low collection cost and lack of competition with other industrial uses, cassava waste is an ideal substrate for the microbial production of value-added products.

3.1. Ethanol

Bioethanol is one of the most important renewable fuels contributing to the reduction in the negative environmental impact generated by the worldwide burning of fossil fuels. The use of ethanol as an alternative fuel has been steadily increasing around the world for several reasons. Domestic production and the use of ethanol for fuel can decrease dependence on fossil fuels, reduce trade deficits and create jobs in rural areas as well as reduce air pollution and carbon dioxide buildup [7].

Currently, ethanol fuel is produced almost exclusively from cornstarch and sucrose from sugarcane. However, a number of different materials can be used to produce ethanol fuel, such as wheat, molasses, grain, cassava and plant fiber, but the technology of bioethanol production using lignocelluloses biomass as raw material remains in an incipient stage. Using food crops such as wheat, corn and grains to produce bioethanol makes the process expensive. Cassava waste is considered a good feedstock for the production of ethanol due to its high starch content and the fact that it is nonfood material with the virtues having large distribution and being both relatively cheap and abundant in many countries.

The effect of combining heat treatment and acid hydrolysis on grated cassava peels has been investigated with regard to ethanol production. The hydrolysate made from 0.3 M H_2SO_4 at 120 °C and 1 atm pressure for 30 min achieved 60% process efficiency in hydrolyzing the cellulose and lignin materials present in grated cassava peels. The ethanol yield was 3.5 (v/v %) after yeast fermentation by *Saccharomyces cerevisiae* [8].

An alternative cassava bagasse saccharification process, which employed the multi-activity enzyme from *Aspergillus niger* BCC17849, was developed for ethanol production by *Candida tropicalis*. The crude multi-enzyme composed of non-starch polysaccharide hydrolyzing enzyme activities, including cellulase, pectinase and hemicellulase, acted cooperatively to release the trapped starch granules from the fibrous cell wall structure for subsequent saccharification through the degrading activity of the raw starch. A high yield of fermentable sugars, equivalent to 716 mg of glucose and 67 mg of xylose/g from cassava bagasse, was obtained after 48 h of incubation at 40 °C and pH 5, using the multi-enzyme, which was greater than the yield obtained from the optimized combinations of the corresponding commercial enzymes. The combined process produced 14.3 g/l ethanol from 4% (w/v) cassava bagasse after 30 h of fermentation. The productivity rate of 0.48 g/l/h is equivalent to 93.7% of the theoretical yield based on total starch and cellulose or 85.4% based on total fermentable sugars. The authors concluded that the multi-enzyme saccharification reaction can be performed simultaneously with the ethanol fermentation process [9].

A novel full recycling process of ethanol production using a two-stage anaerobic treatment has been proposed for the treatment of distillery wastewater. The anaerobic digestion liquid was recycled for ethanol fermentation, thereby establishing a full recycling process. In this process, cassava was used as the raw material to produce bioethanol and the anaerobic digestion step, which has a low operation cost, removed most of the impurities from the wastewater while producing a large amount of biogas that can be used to produce electricity, with the wastewater reused in the production of bioethanol. Such an operation can result in zero wastewater discharge and little energy consumption [10]. Cassava waste has been used for ethanol production employing different processes. Table 2 displays different processes and microorganisms cultivated on cassava waste for the production of ethanol.

Table 2. Microorganisms and processes involved in ethanol production using cassava waste

Microorganism	Process	Cassava waste	Ref
Candida tropicalis	Enzymatic hydrolysis	Bagasse	[9]
Saccharomyces cerevisiae	Enzymatic hydrolysis	Bagasse	[11]
S. cerevisiae	Heat and acid hydrolysis	Peels	[8]
S. cerevisiae	Acid hydrolysis	Bagasse	[12]
S. cerevisiae	Enzymatic hydrolysis	Bagasse	[13]
S. cerevisiae	Anaerobic treatment	Wastewater	[10]

3.2. Organic Acids

Organic acids are among the most important products stemming from microbial cultivation on cassava waste. Considerable attention has recently been paid to the production of organic acids from different agro-industrial residues.

Citric acid production has been well studied and is used in different industrial processes, such as in the food, pharmaceutical, cosmetic and plastic industries [14]. Vandenbergue et al. [15] evaluated three different agro-industrial wastes (sugar cane bagasse, coffee husk and cassava bagasse) with regard to their efficiency in the production of citric acid through the cultivation of *Aspergillus niger* employing solid-state fermentation. Among the substrates, cassava bagasse led to the best growth and highest yield of citric acid. The results demonstrated that the fungal strain exhibited good adaptation to the substrate (cassava bagasse) and increased the protein content (23 g/kg) in the fermented matter. Citric acid production reached a maximum (88-g/kg dry matter) when fermentation was carried out with cassava bagasse with an initial moisture of 62% at 26 °C for 120 h. Prado et al. [16] tested the effect of different percentages of gelatinized starch from cassava bagasse on the production of citric acid using solid-state fermentation in a horizontal drum and tray-type bioreactors. Gelatinization was employed to make the starch structure more susceptible to consumption by the fungus. The best results

(26.9 g/100g of dry cassava bagasse) were obtained in the horizontal drum bioreactor using 100% gelatinized cassava bagasse. The authors concluded that the tray-type bioreactor offers advantages and shows promise for large-scale citric acid production in terms of processing costs.

Fumaric acid ($C_4H_4O_4$) has a wide range of uses. As it possesses a double bond and two carboxylic groups, it is an interesting intermediate in chemical synthesis involving esterification reactions. It is also used as an acidulant in the food, beverage and pharmaceutical industries due to its non-toxic, non-hygroscopic properties. Most fumaric acid currently used is produced through the catalytic oxidation of benzene. As benzene is a well-known carcinogenic substance, there is a need for new routes for producing fumaric acid for the food and pharmaceutical industries and several trials have been carried out involving fermentation with fungi and renewable resources [17]. The enzymatic hydrolysate of cassava bagasse has been used as the sole carbon source to produce fumaric acid by submerged fermentation using different strains of *Rhizopus*. Six different sources of nitrogen and six different compositions of the enzymatic hydrolysate were used. The strain *Rhizopus formosa* MUCL 28422 was selected as the best fumaric acid producer, yielding 21.28 g/1 in a medium containing cassava bagasse as the sole carbon source [18].

3.3. Aroma and Flavor Compounds

The synthesis of flavor and fragrance compounds using biotechnological processes has played an increasing role in the food, cosmetic, chemical and pharmaceutical industries due to an increasing preference on the part of consumers for natural food additives and other compounds of biological origin. Microorganisms can produce aroma and flavor compounds during the fermentation of certain foods and beverages, such as cheese and yogurt. The microbial generation of specific aroma compounds through fermentation and later extraction has also been reported [19, 20].

Monosodium glutamate (MSG) is a popular flavor enhancer and food additive. Solid residue from cassava starch factories could serve as a low-cost substrate for glutamic acid production, which is the precursor of monosodium glutamate, employing submerged fermentation using *Brevibacterium divaricatum*. With 0.7% ammonium nitrate in the medium, the highest glutamate yield of 3.86% based on the weight of the residues was obtained at

30 °C at a pH of 7.0. The highest glutamate production was recorded with a 5% inoculum size at an agitation speed of 180 rpm [21].

Under cassava wastewater cultivation, *Geotrichum fragrans* produced fruit aroma volatile compounds. The following volatile compounds were identified in the cassava liquid waste after 72 h of cultivation: 1-butanol, 3-methyl 1-butanol (isoamylic alcohol), 2-methyl 1-butanol, 1-3 butanodiol and phenylethanol, ethyl acetate, ethyl propionate, 2-methyl ethyl propionate and 2-methyl propanoic. Substrate supplementation with glucose (50 g/l), fructose (50 g/l) and aqueous yeast extract (200 ml/l) did not affect the qualitative and quantitative profiles of the volatile compounds [20]. Five agro-industrial residues were evaluated as substrate for cultivating a strain of *Kluyveromyces marxianus* for the production of aroma compounds in solid-state fermentation. The results proved the feasibility of using cassava bagasse and giant palm bran (*Opuntia ficus indica*) as substrates to produce fruity aroma compounds by the yeast culture. Head-space analysis of the culture using gas chromatography revealed the production of nine and eleven compounds from palm bran and cassava bagasse, respectively, which included alcohols, esters and aldehyde. Ethanol was the compound produced at the highest concentration when using palm bran and ethyl acetate was produced at the highest concentration when using cassava bagasse [22].

The use of two agro-residues (liquid cassava waste and orange essential oil) in the biotransformation of R-(+)-limonene to R-(+)-α-terpineol was investigated employing *Fusarium oxysporum*. R-(+)-α-terpineol has a floral, typically lilac odor and one of the most commonly used fragrance compounds. Cassava wastewater proved to be a suitable substrate for mycelia growth, leading to good, rapid growth. *F. oxysporum* converted R-(+)-limonene to R-(+)-α-terpineol, yielding nearly 450 mg/l after three days of transformation [23].

3.4. Biomass Production

Cassava waste has also been used for biomass production by fermentation. Solid cassava waste has been used extensively. Cassava peel was readily degraded and used by a strain of *Rhizopus* growing in solid-state fermentation. Growth was maximal at 45 °C and was proportional to the degree of hydrolysis of the peel. The biomass yield, as weight of dry mycelium from the reducing sugars from the peel, was 51%. After 72 h of fermentation, the peel protein content increased from 5.6% to 16%. The results suggest a possible economic value of cassava peel in the production of fungal biomass and feedstock [24]. Mash prepared from cassava peels was inoculated with either *S. cerevisiae* or *C. tropicalis* and left to ferment for seven days. The results demonstrated that it is possible improve the crude protein content of cassava peel mash with microbial proteins [25].

Cassava bagasse has also been used for mushroom cultivation in solid-state fermentation. Beux et al. [26] compared the cultivation of *Lentinus edodes* on cassava bagasse and sugarcane bagasse individually and together. Both substrates were suitable for mushroom production, with the best results achieved when employing a blend of cassava bagasse (80%) and sugarcane bagasse (20%). The results were reported to be useful in providing a novel alternative technology for shiitake production. Cassava bagasse and sugarcane bagasse have also been compared for mushroom production [27]. Cassava bagasse demonstrated good potential for mushroom cultivation, but the best results were obtained when cassava bagasse was blended with sugarcane bagasse [27].

A semi-solid fermentation technique has been used to convert cassava waste to phosphate biofertilizer by *Aspergillus fumigatus* and *A. niger*. The medium for the semi-solid fermentation was composed of 1% raw cassava starch and 3% poultry droppings as nutrients and 96% ground (0.5 to 1.5 mm) dried cassava peels as the carrier material. At the end of 14 days of fermentation, the medium converted to biofertilizer was considerably modified in texture, becoming less coarse to the touch and darker in color; moreover, the product remained shelf stable for the entire five months it was stored [28].

3.5. Antibiotics

Submerged cultures and culture media are usually used to produce antibiotics, but culture conditions affect the kind and quantity of antibiotic production. Solid-state fermentation gives a product that is more stable than that of submerged culture and requires less energy input [29]. Cellulosic materials, such as solid cassava waste, are abundantly available globally and can be used as substrate for the production of antibiotics through solid-state fermentation.

Tetracyclines are broad-spectrum antibiotics that are useful in a variety of infections caused by bacteria, rickettsias, trachoma, coccidia, balantidium and mycoplasma. Asagbra et al. [29] used *Streptomyces spp.* to produce tetracycline from different agricultural wastes. Peanut (groundnut) shells were the most effective substrate (4.36 mg/g) followed by corncob (2.64 mg/g), cassava peels (2.16 mg/g) and corn pomace (1.99 mg/g).

3.6. Methane and Hydrogen Gas

Solid cassava waste has 50 to 60% starch in dry matter and has considerable potential as a raw material for the production of biogas. However, low concentrations of nutrients such as nitrogen and the low buffering capacity of this waste constitute a limitation to the conversion into biogas. Anaerobic co-digestion is a promising technology widely applied to many waste treatments, especially pig and cattle manure. Manure is an excellent co-substrate due to its high buffering capacity and the fact that it is rich in a wide variety of nutrients necessary for optimal bacterial growth [30, 31]. Panichnumsin et al. [32] determined the potential of the co-digestion of cassava bagasse with pig manure in an anaerobic digestion process as a raw material substrate for the production of methane. Co-digestion resulted in greater methane production and a reduction in volatile solids, but a lower buffering capacity. Compared to the digestion of pig manure alone, the specific methane yield increased 41% when co-digested with cassava bagasse at concentrations of up to 60% of the volatile solids. However, the high level of the easily degradable fraction in the feedstock affected reactor stability, especially when the reactor was fed with feedstock containing an inappropriate C:N ratio. Co-digestion with pig manure helps increase the buffering capacity and provides a nitrogen source for microbial synthesis resulting in a stable anaerobic digestion process.

Bio-hydrogen production from renewable sources, which is considered a "green technology", has received attention in recent years as an approach to promoting the sustainable development of global prosperity. Hydrogen can be produced from different raw materials, including fossil fuels, biomass and water. Hydrogen production from biological processes is of considerable interest, as such processes can be operated at room temperature and pressure, which requires less energy consumption and is more environmentally friendly [33]. Zong et al. [34] employed a two-step process of sequential anaerobic (dark) and photo-heterotrophic fermentation to produce hydrogen from cassava and food waste. In dark fermentation, the average yield of hydrogen was approximately 199 ml H_2/g of cassava and 220 ml H_2/g of food waste. In the subsequent photo-fermentation, the average yield of hydrogen from the dark fermentation effluent was approximately 611 ml H_2/g of cassava and 451 ml H_2/g of food waste. The total hydrogen yield in the two-step process was estimated at 810 ml H_2/g of cassava and 671 ml H_2/g of food waste. The results demonstrate that cassava and food waste were both ideal biomass resources for bio-hydrogen production. The hydrogen-producing process also removed 84.3% of the COD from the cassava and 80.2% of the COD from the food waste in the fermentation batches, suggesting that it also promises to be useful in waste treatment.

3.7. Xanthan Gum

Xanthan gum is the common name of a complex microbial exopoly-saccharide produced from a fermentative process using *Xanthomonas sp.* Due to its physical properties, xanthan gum is used as a stabilizing agent in a wide range of emulsions, suspensions and foam products. This polymer has been increasingly applied as a thickening, stabilizing and gelling agent in a large range of industries as well as in food products [35]. Woiciechowski et al. [36] hydrolyzed cassava bagasse using HCl and the hydrolysate was used for the production of xanthan gum using a bacterial culture with *X. campestris*. Cassava bagasse hydrolysate supplemented with a nitrogen source was considered a suitable substrate. Maximal xanthan gum yield (about 14 g/L) was produced when the medium was supplemented with potassium nitrate and fermented for 72 h.

3.8. Amylase

Amylase is used in the brewing and bakery industries as well as in the industrial manufacturing of glucose, invert sugars and high fructose syrup. For the production of amylase by microorganisms, a commercially available soluble starch or other sugar, such as maltose, lactose and galactose, is used [37]. Soluble starch and cassava peels were compared for the extracellular production of amylase using *Aspergillus flavus* and *Aspergillus niger*. The greatest amylase activity was obtained with *A. flavus* when grown on cassava peels and was 170-fold greater than that using the soluble starch, while that of *A. niger* was 16-fold greater. The results demonstrate that cassava peel may be a better substrate for the production of amylase by *A. flavus and A. niger* than commercial soluble starch [37].

3.9. Lactic Acid

Lactic acid has a wide range of applications in the pharmaceutical, food, leather, textile and cosmetic industries. The demand for lactic acid has recently increased as a result of the environmentally friendly nature of its polymer – polylactic acid. Lactic acid is used to improve the physical properties of garbage bags, agricultural plastic sheeting and computer parts. Due to its biocompatible and bioabsorbable characteristics, it is also used in sutures and surgical implants [38]. Lactic acid is commercially produced through chemical synthesis and fermentation. Refined sugars, such as glucose or sucrose, have been the most frequently used carbon sources and yeast extract has been the most frequently used nitrogen source for lactic acid production. However, this is economically unfavorable and a cheaper starchy substrate can replace refined sugars [39].

The saccharification and fermentation of cassava bagasse has been carried out in a single step for the production of L-(+)-lactic acid by *Lactobacillus casei* and *Lactobacillus delbrueckii*. Using 15.5% w/v of cassava bagasse as the raw material, 96% maximal conversion of starch to lactic acid was obtained with *L. casei,* with a productivity rate of 1.40 mg/mL/h and maximal yield of 83.8 mg/mL. The authors concluded that a simultaneous cassava bagasse saccharification and fermentation process is a cost-effective and eco-friendly method for converting cassava starch to L-lactic acid, thereby aggregating value to cassava bagasse [40].

Coelho et al. [39] investigated the effects of different medium components on cassava wastewater for the production of L(+)-lactic acid by *Lactobacillus rhamnosus* B103. The response surface method was used to identify the medium components with a significant effect on lactic acid production. The supplementation of Tween 80 and corn steep liquor to cassava wastewater at concentrations (v/v) of 1.27 ml/l and 65.4 ml/l, respectively, led to a lactic acid concentration of 41.65 g/l after 48 h of fermentation. Maximal lactic acid concentration produced in the reactor after 36 h of fermentation was 39.0 g/l using the same medium.

3.10. Biosurfactants

Biosurfactants constitute a diverse group of surface active molecules synthesized by microorganisms. The most important advantage of bio-surfactants when compared to synthetic surfactants is their ecological acceptance due to their low toxicity and biodegradable nature. Biosurfactants are used in a wide variety of applications in the food, cosmetic, pharmaceutical and chemical industries. The main factor hindering the widespread use of biosurfactants is the economics of their production. A reduction in the production costs of microbe-derived surfactants requires enhanced efficiency of the biosynthesis process and the selection of inexpensive medium components. To this end, agro-industrial byproducts and wastes seem to be good components [41].

Nitschke and Pastore [42] investigated the production and properties of a biosurfactant synthesized by the LB5a strain of *Bacillus subtilis* using cassava wastewater as the substrate. The microorganism was able to grow and produce surfactant on cassava waste, reducing the surface tension of the medium to 26.6 mN/m and obtaining a crude surfactant concentration of 3.0 g/l after 48 h. The authors concluded that cassava wastewater proved to be a suitable substrate for biosurfactant biosynthesis, providing not only bacterial growth and product accumulation, but also a surfactant with interesting and useful properties.

Cassava wastewater, cooking oil waste and a combination of the two were evaluated as alternative low-cost carbon substrates for the production of rhamnolipids and polyhydroxyalkanoates (PHAs) by various strains of *Pseudomonas aeruginosa*. The best overall production of rhamnolipids and PHAs was obtained with cassava wastewater plus frying oil waste as the carbon source, with PHA production corresponding to 39% of the cell dry weight and rhamnolipid production corresponding to 660 mg/l. The rhamnolipids had similar or better tensioactive properties when compared with the commercial rhamnolipid mixture JBR599 (Jeneil Biosurfactant Co., Saukville, USA). The use of cassava wastewater plus cooking oil waste as substrate for the simultaneous production of both biomolecules can contribute toward a reduction in production costs and environmental problems related to waste discharge as well as an increase in the economic value of these residues [6, 43].

CONCLUSION

Cassava processing is generally considered to contribute significantly to environmental pollution due the large amounts of waste produced. The biotransformation of cassava waste through microbial conversion can help solve this environmental problem, increasing the economic value of these residues and could be economically useful for the production of value-added products. The process involving the biotransformation on cassava waste is carried out with either liquid fermentation or solid-state fermentation. In both processes, cassava waste is easily degraded by microorganisms without any pretreatment or going through a saccharification, acid hydrolysis or enzymatic process, while obtaining a high yield of fermentable sugars.

Most cassava processing plants are small or medium scale and the process of value-added products should be relatively simple, without the involvement of sophisticated equipment. Moreover, it is important to determine the optimal medium and nutrient compositions that favor value-added production. This generates new analytical and methodological difficulties that demand the development of specific methods based on the waste used, thus reinforcing the need for a multidisciplinary approach combining chemical, biological and engineering technologies.

The main factor hindering the widespread use of microbe-derived products is the economics of their production. A reduction in the production costs requires enhanced efficiency of the biosynthesis process and the selection of inexpensive medium components. Research should focus on the screening of microorganisms with high conversion efficiency regarding enzymes, protein and secondary metabolites from waste carbohydrates, with an increase in the employment of cassava waste and consequent contribution to the widespread use of microbe-derived products.

In conclusion, waste from the cassava processing industry shows potential as a good substrate for the microbial conversion of value-added bio-products, such as biofuels, biochemicals and biomaterials.

REFERENCES

[1]   FAO Food agricultural organization—statistical—database. 2005. Food and Agriculture Organization of the United Nations, Rome. Available from: http://www.fao.org.
[2]   FAO Strategic environmental assessment. Proceedings of the validation forum on the global cassava development strategy. 2000. Food and Agriculture Organization of the United Nations, Rome. Available from: http://www.fao.org.
[3]   Ubalua, AO. Cassava wastes: treatment options and value addition alternatives. *African Journal of Biotechnology*. 2007 6, 2065-2073.

[4]   Jonh, RP. Biotechnological potential of cassava bagasse. In: Singh nee'Nigam, P; Pandey, A, editors. Biotechnology for agro-industrial residues utilization. London. *Springer*; 2009; 225-236.

[5]   Cereda, MP. Resíduos da industrialização da mandioca no Brasil. São Paulo: *Paulicéia*; 1994.

[6]   Costa, SGVAO; Lépine, F; Milot, S; Déziel, E; Nitschke, M; Contiero, J. Cassava wastewater as substrate for the simultaneous production of rhamnolipid and polyxydroxyalkanoates by *Pseudomonas aeruginosa*. *J. Ind. Microbiol. Biotechnol.* 2009 36, 1063–1072.

[7]   Badger PC. Ethanol from cellulose: a general review. In: Janick, J; Whipkey A, editors. Trends in new crops and new uses. Alexandria. ASHS Press; 2002; 17–21.

[8]   Agu, RC; Amadife, AE; Ude, CM; Onyia, A; Ogu, EO; Okafor, M; Ezejiofor, E. Combined heat treatment and acid hydrolysis of cassava grate waste (CGW) biomass for ethanol production. *Waste Management* 1997 17, 91-96.

[9]   Rattanachomrsi, U; Tanapongpipat, S; Eurwilaichitr, L; Champreda, V. Simultaneous non-thermal saccharification of cassava pulp by multi-enzyme activity and ethanol fermentation by *Candida tropicalis*. *Journal of Bioscience and Bioengineering* 2009 107, 488-493.

[10]   Zhang, QH; Lu, X; Tang, L; Mao, ZG; Zhang, JH; Zhang, HJ; Sun, FB. A novel full recycling process through two-stage anaerobic treatment of distillery wastewater for bioethanol production from cassava. *Journal of Hazardous Material* 2010 179, 635-641.

[11]   Teerapatr, S; Chollada, S; Bongotrat, P; Wichien, K; Sirintip, C. (2004) Utilization of waste from cassava starch plant for ethanol production. *Joint International Conference on "Sustainable Energy and Environment* (SEE)", Hua Hin, Thailand.

[12]   Martin, C; Lopez, Y; Plasencia, Y; Hernandez, E. Characterisation of agricultural and agro-industrial residues as raw materials for ethanol production. *Chem. Biochem. Eng. Q* 2006 20, 443-447.

[13]   Ray, RC; Mohapatra, S; Panda, S; Kar, S. Solid substrate fermentation of cassava fibrous residue for production of α-amylase, lactic acid and ethanol. *Journal of Environmental Biology* 2008 29, 111-115.

[14]   Pandey, A; Soccol, CR; Nigam, P; Soccol, VT; Vandenberghe, LPS; Mohan, R. Biotechnological potential of agro-industrial residues. II: cassava bagasse. *Bioresource Technology*. 2000 74, 81-87.

[15] Vandenberghe, LPS; Soccol, CR; Pandey, A; Lebeault, JM. Solid-state fermentation for the synthesis of citric acid by *Aspergillus niger*. *Bioresource Technology*, 2000 74, 175-178.

[16] Prado, LC; Vandenberghe, LPS; Woiciechowski, AL; Rdrigues-Leon, JA; Soccol, CR. Citric acid production by solid-state fermentation on a semi-pilot scale using different percentages of treated cassava bagasse. *Brazilian Journal of Chemical Engineering* 2005 22, 547-555.

[17] Roa Engel, CA; Straathof, AJJ; Zijlmans, TW; van Gulik, WM; van der Wielen, LAM. Fumaric acid production by fermentation. *Appl. Microbiol. Biotechnol.* 2008 78, 379-389.

[18] Carta, FS; Soccol, CR; Ramos, LP; Fontana, JD. Prouction of fumaric acid by fermentation of enzymatic hydrolysates derived from cassava bagasse. *Bioresource Technology* 1999 68, 23-28.

[19] Bramorski, A; Christen, P; Ramirez, M; Soccol, CR; Revah, S. Production of volatile compounds by the edible fungus *Rhizopus oryzae* during solid state cultivation on tropical agroindustrial substrates. *Biotechnol. Lett.* 1998 20, 359-362.

[20] Damasceno, S; Cereda, MP; Pastore, GM; Oliveira, JG. Production of volatile compounds by *Geotrichum fragans* using cassava wastewater as substrate. *Process Biochemistry* 2003 39, 411-414.

[21] Jyothi, AN; Sasikiran, K; Nambisan, B. Balagopalan, C. Optimisation of glutamic acid production from cassava starch factory residues using *Brevidobacterium divaricatum*. *Process Biochemistry* 2005 40, 3576-3579.

[22] Medeiros, ABP; Pandey, A; Freitas, RJS; Christen, P; Soccol, CR. Optimization of the production of aroma compounds by *Kluyveromyces marxianus* in solid-state fermentation using factorial design and response surface methodology. *Biochemical Engineering Journal* 2000 6, 33-39.

[23] Marostica Jr, MR; Pastore, GM. Production of R-(+)-α-terpineol by the transformation of limonene from orange essential oil, using cassava wastewater as medium. *Food Chemistry* 2007 101, 345-350.

[24] Ofuya, CO; Nwajiuba, CJ. Microbial degradation and utilization of cassava peel. *World Jounal of Microbiology and Biotechnology* 1990 6, 144-148.

[25] Antai, SP; Mbongo, PM. Utilization of cassava peels as substrate for crude protein formation. *Plants Food for Human Nutrition* 1994 46, 345-351.

[26] Beux, MR; Soccol, CR; Marin, B; Tonial, T; Roussos, S. Cultivation of *Lentinus edodes* on the mixture of cassava bagasse and sugarcane

bagasse. In: Roussos, S., Lonsane, B.K., Raimbault, M., Viniegra-Gonzalez, G. editors. *Advances in Solid State Fermentation*. Dordrecht. Kluwer Academic Publishers; 1995; 499-511.

[27] Barbosa, MCS; Soccol, CR; Marin, B; Todeschini, ML; Tonial, T; Flores, V. Prospect for production of *Pleurotus sajor-caju* from cassava fibrous waste. In: Roussos, S., Lonsane, B.K., Raimbault, M., Viniegra-Gonzalez, G. editors. *Advances in Solid State Fermentation*. Dordrecht. Kluwer Academic Publishers; 1995; 513-525.

[28] Ogbo, FC. Conversion of cassava wastes for biofertilizer production using phosphate solubilizing fungi. *Bioresource Technology* 2010 101, 4120-4124.

[29] Asagbra, AE; Sanni, AI; Oyewole, OB. Solid-state fermentation production of tetracycline by *Streptomyces* strains using some agricultural wastes as substrate. *Would Journal of Microbiology and Biotechnology* 2005 21, 107-114.

[30] Hartmann, H; Angelidaki, I; Ahring, BK. Co-digestion of the organic fraction of municipal waste. In: Mata-Alvarez, J, editor. Biomethanization of the organic fraction of municipal solid waste. London. IWA Publishing; 2002; 181–200.

[31] Ward, AJ; Hobbs, PJ; Holliman, PJ; Jones, DL. Optimisation of the anaerobic digestion of agricultural resources. *Bioresource Technology* 2008 99, 7928–7940.

[32] Panichnumsin, P; Nopharatana, A; Ahring, B; Chaiprasert, P. Production of methane by co-digestion of cassava pulp with various concentrations of pig manure. *Biomass and Bioenergy* 2010 34, 1117-1124.

[33] Li, CL; Fang, HHP. Fermentative hydrogen production from wastewater and solid wastes by mixed cultures. *Crit. Rev. Environ. Sci. Technol.* 2007 37, 1–39.

[34] Zong, W; Yu, R; Zhang, P; Fan, M; Zhou, Z. Efficient hydrogen gas production from cassava and food waste by a two-step process of dark fermentation and photo-fermentation. *Biomass and Bioenergy* 2009 33, 1458-1563.

[35] Casas, JA; Santos, VE; Garcia-Ochoa, F. Xanthan gum production under several operational conditions: molecular structure and rheological properties. *Enzyme Microb. Technol.* 2000 26, 282–291.

[36] Woiciechowski, AL; Soccol, CR, Rocha, SN; Pandey, A. Xanthan gum production from cassava bagasse hydrolysate with *Xanthomonas campestri* using alternatives sources of nitrogen. *Applied Biochemistry and Biotechnology* 2004 118, 305-312.

[37] Sani, A; Awe, FA; Akinyanju, JA. Amylase synthesis in *Aspergillus flavus* and *Aspergillus niger* grown on cassava peel. *Journal of Industrial Microbiology* 1992 10, 55-59.

[38] Datta, R; Tsai, SP; Bonsignor, P; Moon, S; Frank, J. Technological and economical potential of polylactic acid and lactic acid derivatives. *FEMS Microbiol. Rev.* 1995 16, 221–231.

[39] Coelho, LF; de Lima, CJB; Bernardo, MP; Alvarez, GM; Contiero, J. Improvement of *L*-(+)-acid lactic production from cassava wastewater from *Lactobacillus rhamnosus* B103. *J. Sci. Food.Agric.* 2010 90, 1944-1950.

[40] John, RP; Madhavan Nampoothiri, K; Pandey, A. Silmutaneous saccharification and fermentation of cassava bagasse of *L*-(+)-acid lactic production using *Lactobacilli*. *Applied Biochemistry and Biotechnology* 2006 134, 263-272.

[41] Nitschke, M; Costa, SGVAO; Contiero, J. Rhamnolipid Surfactants: An update on the general aspects of these remarkable biomolecules. *Biotechnology Progress* 2005 21, 1593-1600.

[42] Nitschke, M; Pastore, GM. Production and properties of a surfactant obtained from *Bacillus subtilis* grown on cassava wastewater. *Bioresource Technology* 2006 97, 336-341.

[43] Costa, SGVAO; Nitschke, M; Lépine, F; Déziel, E; Contiero J. Structure, properties and applications of rhamnolipids produced by *Pseudomonas aeruginosa* L2-1 from cassava wastewater. *Process Biochemistry* 2010 45, 1511-1516.

In: Cassava: Farming, Uses, and Economic Impact   ISBN:978-1-61209-655-1
Editor: Colleen M. Pace   © 2012 Nova Science Publishers, Inc.

Chapter 5

USE OF CASSAVA STARCH FOR EDIBLE FILMS AND COATINGS FORMULATION

Lía N. Gerschenson and Silvia K. Flores[*]

Industry Department, School of Natural and Exact Sciences (FCEN),
Buenos Aires University (UBA).Ciudad Universitaria,
Ciudad Autonoma de Buenos Aires, Argentina
Members of the National Scientific and
Technological Research Council-Argentina (CONICET)

ABSTRACT

Cassava is produced in Latin America, Asia and Southern Africa. In Latin America, cassava is popularly used as a meal, as animal fodder or cooked and eaten as a vegetable and part of its production is exported. It has been seen that cassava starch is used to a much lesser extent than other starches, like corn one, in food industry. Anyhow, its importance as a source of starch is growing rapidly, especially because its price in the world market is low when compared to starches from other sources.

Edible films and coatings are not designed for totally replacing traditional packaging but to help, as an additional stress factor, for protecting food products, improving quality and shelf life without

[*] Corresponding author: Silvia Karina Flores e-mail: sflores@di.fcen.uba.ar Industry Department, School of Natural and Exact Sciences (FCEN), Buenos Aires University (UBA).Ciudad Universitaria. Intendente Güiraldes 2620, (1428) Ciudad Autonoma de Buenos Aires, Argentina.Phone: 54–1 –4576-3366/3397. Fax number: 54–11–4576-3366.

impairing consumer acceptability. They can control moisture, gases, lipid migration and can also be carriers of additives and nutrients. Cellulose, gums, starch and proteins have been used to formulate edible films and plasticizers are usually employed to enhance its mechanical properties.

The objective of this chapter is to analyze the use of cassava starch for edible films and coatings formulation. The barrier and mechanical characteristics of these edibles according to production technique and formulation as well as their effectiveness for supporting different anti-microbials is considered. It is also revised the possibility of their application to food products.

INTRODUCTION

Starch

Between food grade carbohydrates, the starch occupies an important place. It is the more important reserve carbohydrate in higher plants and is considered the second more abundant natural biopolymer after cellulose. It is amply distributed in the vegetal domain. According to Liu (2005) the average starch content (%, dry basis) of some food grade vegetable sources is: 67 % for wheat, 57% for corn, 89% for white rice, 75% for potatoes and 90% for sweet potato or cassava. It is the basic source of energy for the major part of the world population. In relation to human nutrition, the starch has an important role for the provision of metabolic energy that allows the body to accomplish its different functions.

The starch is one of the more important raw materials for the industry. In its native state, starch has limited applications but chemical and physical modifications allow its use in a great variety of industries including food, paper, textile and plastic industries.

Starch is constituted mainly by a mixture of two biopolymers (glucanes): a linear fraction, the amylose and a fraction highly branched, the amylopectin. The starch is stored in the plant organs as discrete particles or granules which are dense, water insoluble and their size is comprised between 1 and 100 μm. A typical granule consists of, approximately, a billion molecules. When granules are heated in the presence of enough water, a phase transition (order to disorder) occurs. Starch gelatinization is the disruption of the order inside the granules. That transition is accompanied by changes like granular swelling, fusion of crystalline regions, lost of birefringence and viscosity development (BeMiller and Whistler, 1996; Zobel, 1994).

The word retrogradation has been developed to describe the physical changes suffered by starch after gelatinization. This process occurs when starch molecules reassociate and form an ordered structure during storage. During retrogradation, molecular interactions occur, principally, hydrogen bridges between starch chains, being these interactions dependent on time and temperature. Retrogradation is important for the industrial use of starch because it might be the desired ending point for some applications but it can also determine the instability of some starchy pastes (BeMiller and Whistler, 1996).

Cassava Starch

Cassava is produced in Latin America, Asia and Southern Africa (FAO, 2004). In the year 2000, the world production of cassava attained an amount of approximately 200 millions of Ton. In Argentina, the production of cassava is a regional activity developed by small producers. It is fundamentally produced in Misiones province and, in a smaller volume in Formosa, Corrientes and Chaco and the cassava starch is the main industrial product obtained from the tuber. In the year 2000, 14% (28000 Ton) of the production of tubers of the region was used by the industry and the national production of starch attained a value of 7100 Ton. That year, there were also imported 2200 Ton and the consumption of cassava starch attained a value of 9289 Ton. That starch is one of the preferred ingredients used for the elaboration of emulsified meat; great starch quantities are used as absorbers and water retaining agents, specially for the sausage industry (De Bernardi, 2002).

Cassava starch, naturally or modified, has some inherent properties that are demanded in the food industry. The preferred properties of cassava starch include: high transparency, determining suitability for developing sauces for ready-to-eat foods; high resistance to acidity, allowing its use for acid-based sauces and jams. It is also applicable for desserts, puddings, soups, fillings and gums due to its high viscosity. It is also adequate for dietetic foods reduced in fat, gluten-free foods and antiallergic diets due to the absence of gluten, phosphates, oils and proteins. As an alternative starch it could replace traditional starches because it is also a lower cost option. Unfortunately it is not easy to replace starches traditionally used because it is difficult to overcome the strong links that exist between producers, starch manufacturers and food industries that utilize this polysaccharide in main importing countries of Europe and North America (FAO, 2004).

Edible Films and Coatings and Their Application for Industrial Preservation

In the last decades, there has been observed a trend to develop preservation methodologies that result less aggressive to the raw material, tending to produce safety foods with better nutritional and organoleptical qualities. Among the proposed technologies there can be find those ones that combine different factors of microbial stress like slight thermal treatment, pH and water activity control, addition of allowed chemical preservatives, use of edible films and coatings, irradiation, tending to optimize the quality of the food (Devece et al., 1999). Edible films and coatings do not pretend to replace completely traditional packaging but are one of the possible factors to be applied for food preservation. Their usefulness is based on the capacity of optimizing global quality, extending shelf life and, possibly, improving the economic efficiency of packaging materials (Kester and Fennema, 1986).

The use of edible films and coatings is related to:

- The interest of the consumers for healthy, high quality, convenient and safety food.
- The collective awareness about the need for protecting the environment and the need for the availability of biodegradable and/or recyclable materials that can contribute to the reduction of the environmental pollution due to the use of synthetic packaging.
- The potential of these films and coatings for improving the appearance and lengthening the shelf life of foods.
- The availability and low cost of the raw materials used for edible films and coatings production.

Interest in edible film and coating development has increased because evidence was obtained about their beneficial effects on fresh and processed foods (Baker et al., 1994; Mei et al., 2002). The functional properties of edibles are: retardation of moisture migration, retardation of gases transportation (O_2, CO_2); retardation of oil and fat migration; retardation of solutes transport; improvement of mechanical properties and structural integrity of food; improvement of the retention of volatile compounds; capacity of acting as supporters and/or carriers of food additives and nutrients (Buonocore et al., 2003; Guilbert, 1988). They can also reduce packaging waste associated with processed foods (Chung et al., 2001; Franssen et al., 2002; Greener-Donhowe and Fennema, 1994). The functional property searched for will depend on the type of food to be coated and its primary mechanism of damage. The marketing of foods is faced with many challenges like producing food of high quality, nutritious, stable and economical and edible films and coatings can help to achieve one or more of these functions (Greener-Donhowe and Fennema, 1994).

Edible films can be prepared from substances recognized as GRAS (generally recognized as safe, according to United States legislation) such as hydrocolloids, lipids and their mixtures. Hydrocolloids (carbohydrates, proteins) can be used when the control of moisture migration is not the target. Lipids are good barriers to water vapor, determining that their combination with hydrocolloids optimize the water barrier properties. The formulation of bi- or multilayer films also attends the need for improving system properties.

Different film forming techniques are reported: coacervation, solvent evaporation, solidification through temperature depression, extrusion (Campos et al., 2010). To insert individual citation into a bibliography in a word-processor, select your preferred citation style below and drag-and-drop it into the document.

For the application of coatings different techniques such as immersion and spraying can be used (Delville et al., 2003; Mc Hugh and Senesi, 2000).

In recent years, active packaging technologies with main emphasis on antimicrobial applications have been especially considered (Suppakul et al., 2003). Particularly, antimicrobial films and coatings have been established as an efficient alternative for controlling food contamination (Chen et al., 1996; Franssen and Krochta, 2003).

The acceptability of the presence of a certain edible film or coating is conditioned by consumer acceptance of the food involved. Sensorial tests have been recently reported for films elaborated on the basis of candelilla wax and whey proteins using a trained panel with experience in milk flavor, transparency, opacity, sweetness and adhesivity (Kim y Ustunol, 2001). Results indicated that films did not have milk flavor, were slightly sweet and adhesive. In another study, small carrots were peeled and coated with a film based on cellulose and they were tested by a trained panel in relation to color changes, white discoloration of the surface and global acceptability concerning appearance under fluorescent light. Also fresh smell and flavor, sweetness, hardness, and crispness were evaluated (Howard, 1996). Coated carrots developed a smaller surface discoloration and obtained a better sensorial score for color intensity, fresh flavor and global acceptability. Chien et al. (2007) evaluated visual appearance, sweetness, sourness and global acceptability of mango slices coated with chitosan based films, during storage at 6°C. Results showed that coated slices were more accepted after seven days of storage while uncoated control resulted non satisfactory; the coating did not impair the natural taste of the slices.

EDIBLE FILMS AND COATINGS FORMULATED ON THE BASIS OF CASSAVA STARCH

Elaboration through Casting: Influence of the Processing Variables and Sorbate Presence on Physico-Chemical and Mechanical Properties

In our laboratory different studies were performed involving the production of edible films on the basis of cassava starch. Flores et al. (2007a) prepared the films using mixtures of cassava starch, glycerol and water (5.0:2.5:92.5, in weight) or of starch, glycerol, potassium sorbate and water (5.0:2.5:0.3:92.2, in weight). Three methods of preparation were assayed:

a) *Method 1:* heating of 300g film forming solution on a magnetic stirrer with hot plate at an initial rate of 1.6°C/min for $\cong$ 25 min, moment at which the system entered in the gelatinization step (gelatinization temperature $\cong$ 70°C). Afterwards, heating was maintained at a lower rate ($\cong$ 0.3°C/min) for an additional period of 40 min. After gelatinization, films were casted over glass plates and dried at 50°C (R.H.: 22 %), for two hours. Drying was completed in a controlled temperature chamber at 25°C and R.H.: 80-90% during a week.

b) *Method 2:* heating of 300g film forming solution on a magnetic stirrer with hot plate at a constant rate of 1.8°C/min for $\cong$ 30 min. In this case, it could be visually appreciated that gelatinization began at $\cong$70°C. After gelatinization, films were casted over glass plates and dried at 50°C (R.H.: 22 %) for two hours. Drying was completed in a chamber at 25°C and R.H.: 80-90% during a week.

c) *Method 3:* heating of 300g film forming solution on a magnetic stirrer with hot plate at a constant rate of 1.8°C/min for $\cong$ 30 min. After gelatinization, films were casted over glass plates and dried at 50°C (R.H.: 22 %) for two hours. Drying was completed over $CaCl_2$ (R.H.: 0%) at 25 °C during two days.

The deformation nature of the films studied at room temperature, under an applied load, was studied with a Dynamic Mechanical Analyzer and was typical of ductile plastics in terms of the stress and strain. As generally occurs for those materials, the films exhibited two characteristic regions of deformation behavior in their tensile stress-strain curves. At strains lower than 10 %, the stress increased rapidly with an increase in the strain and the initial slopes were steep in the elastic region, indicating the high elastic modulus of the materials. At strains higher than 10%, the films showed a slow increase in stress with strain. As can be observed in Table 1, method 1 and method 2 produced films with significantly higher tensile stress (σ) than method 3 for both formulations assayed. Arvanitoyannis et al. (1998), assayed low and high temperature drying methods with hydroxypropyl starch-gelatin films plasticized by polyols, and suggested that at high temperature of drying, polymer chains are trapped in a disordered and entangled state described by a low crystallinity and lower tensile strength. It has been also suggested a more efficient separation of amylose from amylopectin during longer heating times which might promote gelation of the linear starch fraction when gel is cooled (Biliaderis, 1994). Leaching of amylose from granule starts gelatinization process, followed by growth and coarsening associations of chains upon double helix formation and then interlinking of aggregates to form a network with consequently further thickening and rigidity development. Density and lifetime of interactions between continuous matrix of amylose and recrystallised structure of amylopectin also influence the mechanical properties of the films. Slow drying provides a longer time for the occurrence of previously mentioned phenomena and, as a consequence, might contribute to obtain a more elastic film and higher tensile stresses when methods 1 or 2 were applied. Table 1 clearly shows that the films without sorbate, showed a higher tensile stress (σ) for a deformation of 70% than the films with sorbate and that behavior was not dependent on the film obtaining method. Method 1 showed the highest tensile stress in films with or without antimicrobial.

As starch based edible films are systems that represent physical gels formed by molecular ordering and subsequent chain aggregation/ crystallization, it is very advantageous using small strain mechanical testing to unravel how changes in composition, polymer structure and temperature-time history affects structural changes and interactions of starch macromolecules. Table 1 shows that films obtained with method 1 developed the highest storage modulus (E') value and with method 3, the smallest ones: long time of gelatinization and drying (method 1) resulted in a more elastic network as a consequence of higher molecular order and/or interactions developed during retrogradation, as was explained above.

Table 1. Effect of gelatinization/drying technique on mechanical properties of cassava starch films. Flores et al. (2007a)

	E′ (MPa)	E″ (MPa)	σ (MPa) At a deformation of 70%
Method 1			
without sorbate	29.0 ± 9.0	7.2 ± 1.3	2.35 ± 0.18
with sorbate	7.6 ± 0.7	3.1 ± 0.4 [b]	0.74 ± 0.12 [c, d]
Method 2			
without sorbate	13 ± 4	3 ± 1 [b]	1.98 ± 0.26
with sorbate	4.3 ± 0.5 [a]	1.7 ± 0.2 [b]	0.57 ± 0.08 [d]
Method 3			
without sorbate	3.2 ± 1.1 [a]	0.9 ± 0.3 [b]	1.0 ± 0.1 [c]
with sorbate	1.3 ± 0.5 [a]	1.0 ± 0.3 [b]	0.16 ± 0.04

Values followed by the same letter are not significantly different (p>0.05).
σ: tensile stress.
E': storage modulus.
E'': loss modulus.

The lower level of organization of polymer chains attained through method 3, could explain the significant (α: 0.05) increase of loss tangent (tan δ) values for films containing the antimicrobial and obtained through that technique (0.78 ± 0.03 vs. 0.410 ± 0.007 and 0.40 ± 0.01; method 3 vs. 1 and 2). The faster evaporation rate involved in method 3 resulted in a less elastic network. It can be also observed in Table 1 that E' values for films without sorbates were always higher than the ones for films with sorbates, for each processing technique assayed, showing a plasticizing effect of sorbates. This can be attributed to the structural modification of starch network when sorbates were incorporated, which determined that under stress, movements of polymer chains were facilitated. For films without sorbate no significant difference was observed in tan δ for different methods applied (tan δ = 0.254 ± 0.004, 0.244 ± 0.005 and 0.280 ± 0.030 for method 1, 2 and 3 respectively).

It is interesting to remark that two way ANOVA showed a significant influence of sorbate concentration and casting method on E', tan δ and σ as well as a significant interaction between those factors in relation to these mechanical parameters.

Table 2 shows the sorbate concentration two weeks after gelatinization. Method 3 showed the highest sorbate content which can be attributed to the shorter gelatinization and drying process applied for its development. Gerschenson and Campos (1995) showed that heat abuse can result in sorbate destruction.

Table 2. Effect of gelatinization/drying technique on sorbate content and water vapor permeability (WVP) of cassava starch edible films (averages and confidence intervals are reported). Flores et al. (2007a)

	Sorbate content (g/100 g, d.b.)[1]	WVP [2] ($\times 10^{10}$ g / Pa s m)
Method 1		
without sorbate		6.3 ± 0.9 [b, c]
with sorbate	4.00 ± 0.28 [a]	6.1 ± 0.8 [c]
Method 2		
without sorbate		8.1 ± 1.9 [b, c]
with sorbate	4.51 ± 0.89 [a]	8.1 ± 1.0 [b]
Method 3		
without sorbate		14.4 ± 1.5 [d]
with sorbate	5.98 ± 0.71	16.1 ± 1.3 [d]

Values followed by the same letter are not significantly different ($p>0.05$).
[1] d.b.: dry basis.
[2] Film thickness was $\cong 0.30$ mm.

As can be seen in Figure 1, film solubility was similar ($\cong 22$ %) for all processing techniques applied in the absence of sorbates but increased significantly when the antimicrobial was present in films based on cassava starch. The presence of antimicrobial produced less organized networks that showed a solubility of $\cong 31$ % for all gelatinization/drying methods assayed.

Water vapor permeability (WVP) values can be observed in Table 2. Method 3 resulted in significantly higher WVP value as the result of a less tight starch network. It is important to remark that this gelatinization/ drying technique practically doubled the WVP value when compared with method 1. Sorbate presence did not affect WVP. A similar trend was reported by Cagri et al. (2001), for a protein based film containing 0.50 and 0.75% sorbic acid. It is well-known that edible films tend to be poor moisture barriers due to abundant hydrophilic groups in biopolymer matrix; however, technique applied to prepare films can help to improve moisture barrier properties. In our case, method 1 proved to give origin to films with the lowest WVP values.

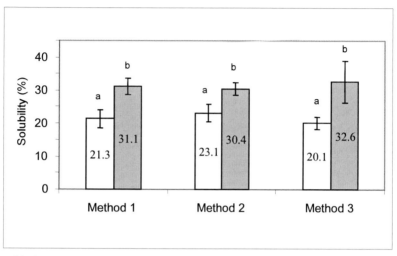

Bars with the same letter are not significantly different (p > 0.05).

Figure 1. Effect of gelatinization/drying technique on solubility of cassava-starch edible films. ■ with sorbates ; □ without sorbates. Averages and confidence intervals are reported. Bars with the same letter are not significantly different (p > 0.05). Flores et al. (2007a).

It is important to remark that two-way analysis of variance (ANOVA) showed that there was a significant interaction between casting technique and sorbate concentration in WVP results: sorbate depressed WVP for method 1 and increased permeability for method 3.

Elaboration through Extrusion. Influence of the Formulation on Physico-Chemical and Mechanical Properties

Flores et al. (2010) used extrusion for preparing films using a mixture of cassava starch and xanthan gum. A three component constrained mixture design with two overall central point replications was used to study physico-chemical and mechanical properties of cassava starch-glycerol based edible films added with xanthan gum (XG) and potassium sorbate (PS). The films included glycerol as plasticizer and studied variables were the concentrations of cassava starch - glycerol blend (S-G. Proportion: 0.89-0.96) with a fixed ratio of 82:18 (starch:glycerol), potassium sorbate (PS. Proportion: 0.01-0.04) and xanthan gum (XG. Proportion: 0.00-0.10). Films obtained were flexible, even, homogeneous and with a slightly variable light brown tone depending on formulation.

Stress at break (σ_b, kPa), strain at break (ε_b, %), elastic modulus (Ec), solubility in water (S) and water vapor permeability (WVP) were evaluated for the different films obtained with a Universal Testing Machine. Mechanical properties were evaluated both in flow and perpendicular directions. For σ_b in flow and in perpendicular direction, S-G and XG presented positive effects while PS showed negative effect, indicating a matrix reinforcing effect of the biopolymers and a plasticizing effect of the preservative. Similar plasticizing behavior of the antimicrobial was reported by Flores et al. (2007a). A synergistic effect between S-G and PS affected the two responses while an antagonic interaction between S-G and XG influenced only σ_b in flow direction. It could be observed that high PS proportions contributed to decrease σ_b value (flow direction) while high XG proportion, increased it. Similar effect but of lower magnitude was observed for σ_b in perpendicular direction.

The ε_b in flow direction showed a positive effect exerted by S-G, PS and XG. PS showed a more pronounced influence. The ε_b, in perpendicular direction was lower as XG concentration increased. A similar trend was reported by Veiga-Santos et al. (2005) who observed a significant and negative effect of the XG on elongation at break of cassava starch based edible films. Such effect could be explained by an interaction between the gum and the starch, which might preclude the occurrence of amylose-amylose interactions. PS contributed to ε_b increase.

Ec in flow and perpendicular directions showed that the effects of S-G and XG were positive and the effect of PS was negative. XG had a greater impact increasing the Ec values, suggesting that the gum imparted a more solid character to the film. Chaisawang et al. (2005) concluded, from dynamic rheological measurements, that interactions occurred between cassava starch and XG resulting in a decrease in the loss tangent as compared with starch alone.

It can be concluded that there were differences in the mechanical behavior according to the direction considered. This anisotropy has been widely reported for synthetic polymers and also, recently, for films based on modified potato starch - glycerol and made by blown technique (Thunwall, 2008). During extrusion process a laminar flow is generated through the die; probably, under such process conditions, biopolymers suffer alignment in the laminar flow field and this can result in the anisotropy observed.

Solubility in water (S) is of major importance since it could condition the actual uses of the films in some technological situations. Results obtained indicated that samples with lower S were obtained for lower PS proportions and higher ratios of S-G/XG.

WVP values ranged from 3.72×10^{-10} to 6.40×10^{-10} g / Pa m s, and were similar to those reported for cassava starch casted edible films (6.1×10^{-10} g / Pa m s) by Flores et al. (2007a) using a similar technique of measurement.

Antimicrobial Edible Films and Coatings Based on Cassava Starch

Nisin Containing Edible Films

Recent studies have focused on the application of edible films on the food surface preventing the diffusion of preservative into the food and inhibiting surface microbial growth (Flores et al., 2007b; Franssen et al., 2002; Ozdemir and Floros, 2001; Sebti et. al, 2004; Vojdani and Torres 1989 a and b). One of the major potentials of this hurdle lies in the storage of semi-moist foods (Chen et al., 1996).

Sanjurjo et al. (2006) prepared films using mixtures of starch, glycerol and water (5.0:2.5:92.5, in weight) or starch, glycerol, nisin and water. In the case of films containing nisin, 15 g of water were replaced by a solution of nisin of a concentration such that each milliliter of final system contained 2000, 3000 or 5000 IU/ml of the antimicrobial (IU: international units of nisin). The pH of the systems was adjusted to 4.0 with citric acid solution (50 %, w/w). Gelatinization was performed at a constant rate of 1.8°C/min for $\cong$ 30 min. It was evaluated the capacity of the films of acting as barriers to antimicrobial contamination exerted by *Lysteria innocua*, through an experiment that used Petri dishes containing TSYE agar with water activity controlled to 0.94 with dextrose (Chirife et al, 1980) and pH 5.0, for resembling a food product. Films were cut in samples of 1 cm diameter resulting in disks with 881, 1322 or 2204 IU/cm^2 of film. The disks were brought in contact with the surface of the agar and then, 10 µl of a culture of *L. innocua* containing 2×10^9 colony forming units per ml (CFU/ml), were dispensed on the disks. Samples were incubated at 28° C during 4 hours and periodically sampled, to test bacterial viability.

It can be observed in Figure 2 a rapid decrease of *L. innocua* viable counts, followed by an additional slow inactivation during the period evaluated. The antimicrobial effect increased with film nisin content, showing a reduction of 4 log cycle during 240 min of contact with a film with 2204

IU/cm^2. On the other hand, films without the natural antimicrobial did not affect initial counts of studied bacteria. Results show the bioavailability of nisin and the usefulness of the antimicrobial containing films for the control of the contamination.

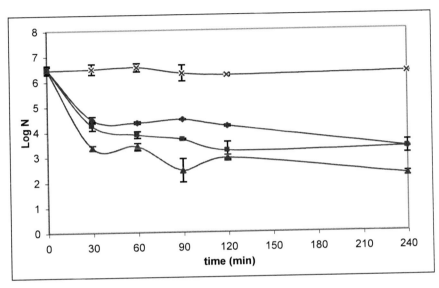

Figure 2. Cassava starch film as barrier to microbial contamination. N: Number of colony forming units per ml. X: Control; ♦: 881 IU /cm^2 ; ■ : 1322 IU /cm^2 ; ▲: 2204 IU /cm^2 . Sanjurjo et al. (2006).

Edible Films with Chitosan

Vasconez et al. (2009) studied edible films based on cassava starch and/or chitosan. Potassium sorbate (PS) was also included in one formulation. Films were obtained by casting technique obtaining stand-alone films. In order to study the performance of the films to prevent microbial contamination of a high water activity (a_w) product, MRS agar or Sabouraud agar with a_w depressed to 0.98 by addition of dextrose (Chirife et al., 1980) and pH adjusted to 4.5 with citric acid 50 % (w/w) were formulated to resemble that kind of products. Inoculums of *Lactobacillus spp* and *Zygosaccharomyces bailii* NRRL 7256 were prepared in MRS (30°C) or Sabouraud (25°C) broth respectively, and incubated until early stationary phase was achieved (24 h).

Disks of 1 cm of diameter were aseptically cut from studied films, weighed and applied on the surface of the MRS or Sabouraud agar plates. Then, 10 µl of the 3-5x10^6 CFU/ml inoculum of *Lactobacillus spp* or *Z. bailii* were seeded on the film disks. Samples were incubated at 30°C (*Lactobacillus*) or 25 °C (*Z.bailii*) for 48 h.

When film effectiveness as antimicrobial barrier was tested against *Lactobacillus spp*, no inhibition was registered since cell growth reached 5×10^8 CFU/g after 24 h of incubation for any formulation.

Regarding chitosan antimicrobial activity, there are contradictory reports in the literature on the base of *in vitro* studies but, in general, susceptibility of bacteria is highly variable being, in particular, lactic acid bacteria less susceptible. On the other hand, yeasts have shown more sensitivity (Devlieghere et al., 2004). Based on mentioned trend, a similar assay was carried out with *Z. bailii*, known deteriorative yeast of acid foods, in order to quantify the film efficiency to prevent cell growth. It can be appreciated (Figure 3) that chitosan-starch based edible films acted as an effective antimicrobial barrier, since a significant (α: 0.05) reduction of yeast population was observed throughout storage in relation to the initial population or to the one observed in a free preservative edible film. Addition of PS to the chitosan- starch edible film did not improve the reduction of growth since no significant differences in yeast population were observed.

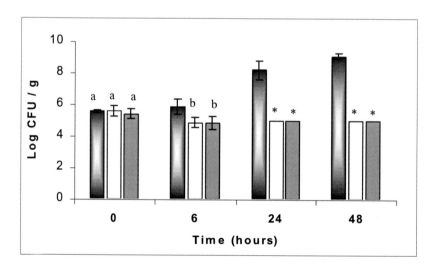

Figure 3. *Z.bailii* growth in the surface of films. Data were obtained from antimicrobial barrier assay. Log CFU / g: log of colony forming units per gram of film. ▣ cassava starch film, ▢ chitosan – cassava starch film, ▨ chitosan – cassava starch – PS film. Columns with the same letter are not significantly different (α: 0.05). * Estimated values. Vasconez et al. (2009).

Edible Coatings with Chitosan

Vasconez et al. (2009) assayed the effect of edible coatings on the shelf life of salmon fillets. Slices of the fish (3 x 3 cm) were cut from the fillets and were then immersed into chitosan or chitosan – cassava starch coating solutions for 3 minutes. Liquid in excess was eliminated and covered slices were allowed to dry in a laminar flow hood at 25°C for 1.5 hours and, then, into a refrigerated chamber at 2°C for 1.5 hours. Pieces of fish were placed on expanded polystyrene trays (3 pieces in each tray), wrapped with PVC film and stored at 2°C for 6 days. Salmon pieces without immersion treatment were used as control systems.

Aerobic mesophilic and psychrophilic bacteria population was enumerated for the surface and, also, for the whole fish muscle, throughout storage, at selected times (0, 3 and 6 days). Figure 4 shows surface muscle count of aerobic mesophilic (panel A) and psychrophilic (panel B) bacteria evaluated from fish coating assay. It can be appreciated that covered fish tended to develop along storage, a lower population of native flora in comparison to control samples. Samples immersed into chitosan coating solution showed cell count reductions of up to 4 and 4.5 cycles log after 3 days of storage, for aerobic mesophilic and psychrophilic bacteria respectively showing, therefore, the highest antimicrobial activity.

Chitosan–cassava starch blend system exhibited lower effectiveness against microorganisms (cell count reductions lower than 2.4 cycles log in contrast with uncoated fish) probably because of the interaction between chitosan and cassava starch which could have reduced the antimicrobial action of chitosan.

The protective action of chitosan on fish tissue had been also observed by others authors (Jeon et al., 2002) who reported that microbial spoilage was retarded when the chitosan coating was applied. However, Devlieghere et al. (2004) studied the effectiveness of chitosan against *Candida lambica* in the presence of others components such as starch, whey protein or NaCl, in media formulation, and established that addition of 30% (w/v) of gelatinized starch to the media, led to a significantly shorter lag phase and a higher growth rate of the yeast.

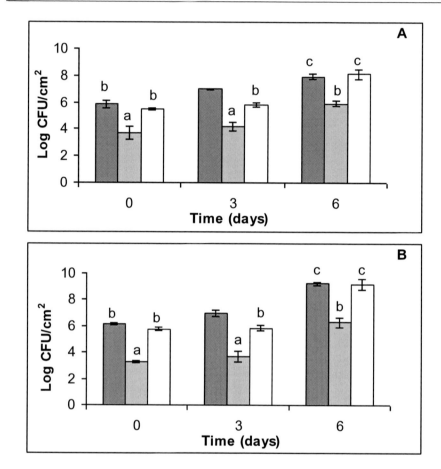

Figure 4. Surface count of aerobic mesophilic (panel A) and psychrophilic bacteria (panel B). Data were obtained along storage of fish. Log CFU / cm^2: log of colony forming units per cm^2 of sample. ■without coating (control), ▨ chitosan coating, □ chitosan – cassava starch coating. Bars with the same letter are not significantly different (α: 0.05). Vasconez et al. (2009).

In general, mesophilic and psychrophilic whole fish muscle populations were higher than surface fish muscle population. This trend suggests that chitosan was more available on the food surface than inside the food muscle.

It must be highlighted that, as storage time increased, coating effectiveness decreased, especially for tissues immersed into chitosan – starch system, where microbial count, after 6 days of incubation, was not, in general, significantly different (α: 0.05) in comparison to control system.

Edible Films with Sorbates

Flores et al. (2010) evaluated the effectiveness of films containing potassium sorbate and fabricated through extrusion, as barriers to yeast contamination. It was observed that extrusion did not affect PS content. For the testing, the films were maintained in contact with an acidified high water activity food model (Sabouraud agar with a_w depressed to 0.98 by addition of glucose and pH adjusted to 4.5 with citric acid) and *Z. bailii* was inoculated at the surface of the film disks.

Figure 5 shows the results obtained. Semisolid food models covered with films fabricated from mixtures with lower PS proportion, such as mixtures containing S-G/PS/XG proportions of 0.99/0.01/0.00 or 0.89/0.01/0.10, suffered a 1.5 or 2.0 log cycle increases of yeast counts after 48 h of storage. On the contrary, films containing higher levels of PS, like 0.89/0.04/0.07 exerted a pronounced antimicrobial action since yeast population remained in the lag phase.

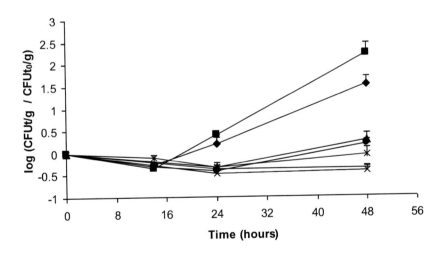

Figure 5. Z.bailii growth in the surface of a film in contact with a semisolid media of aw 0.98 and pH 4.5. S-G/PS/XG proportions: ◆ 0.99/0.01/0.00; ■ 0.89/0.01/0.10; ▲ 0.96/0.04/0.00; × 0.89/0.04/0.07; ✳ 0.975/0.025/0.00; | 0.925/0.04/0.035; • 0.9325/0.025/0.0425. CFU/g: colony forming units per g. Growth was evaluated al each time, t or at initial time, t0. Vertical bars represent standard deviation of the mean (n = 3). Flores et al. (2010).

This assay demonstrated that sorbates were bioavailable to act as a barrier for external yeast contamination. It was observed that XG had a negative effect on PS antimicrobial action as, in general, cell counts were higher when gum was present for mixtures with a similar PS content. This effect was more evident for low preservative proportions. According to results obtained, studied films could have a potential application as an active packaging in relation with its protective action to prevent microbial spoilage.

CONCLUSION

Cassava starch is adequate for formulating edible films and coatings that act as an additional stress factor, for protecting food products. Mechanical properties are adequate but are affected by system formulation and production technique.

The use of cassava starch as the basic polysaccharide for edible films and coatings formulation gives and alternative application for this starch and can help to increase its added value.

ACKNOWLEDGMENTS

The authors acknowledge the financial support of Buenos Aires University (UBA- EX 089), National Scientific and Technological Research Council-Argentina (CONICET-11220090100531), National Agency of Scientific and Technological Promotion (ANPCyT-PICT 38239-PICT 2131).

REFERENCES

Arvanitoyannis, S.I., Nakayama, A. and Aiba, S. (1998). Edible films made from hydoxypropyl starch and gelatin and plasticized by polyols and water. *Carbohydrate Polymers,* 36, 105-119.

Baker, R., Baldwin, E. and Nisperos-Carriedo, M. (1994). Edible coatings and films for processed foods. In J. M. Krochta, E. A. Baldwin and M. O. Nisperos-Carriedo, *Edible coatings and films to improve food quality* (pp 89-104). Lancaster, Pennsylvania: Technomic Publishing Co., Inc.

BeMiller, J.N. and Whistler, R.L. (1996). Carbohydrates. In O.R. Fennema, *Food Chemistry* (pp.157-320). New York: Marcel Dekker Inc.

Biliaredis, C. G. (1994). Characterization of starch networPS by small strain dynamic rheometry. In R. J. Alexander and H. F. Zobel, *Developments in carbohydrate chemistry* (pp. 87-135). St Paul, Minnesota: The American Association of Cereal Chemists.

Buonocore, G.G., Del Nobile, M.A., Panizza, A., Bove S., Battaglia G.and Nicolais L. (2003). Modeling the lysozyme release kinetics from antimicrobial films intended for food packaging applications. *Journal of Food Science,* 68(4), 1365-1370.

Cagri, A., Ustunol, Z. and Ryser, E.T. (2001). Antimicrobial, mechanical, and moisture barrier properties of low pH whey protein-based edible films containing p-aminobenzoic or sorbic acids. *Journal of Food Science,* 66(6), 865-870.

Campos, C., Gerschenson, L.N. and Flores, S.K. (2010). Development of Edible Films and Coatings with Antimicrobial Activity. *Food and Bioprocess Technology* (Published on-line: September 21, 2010). doi:10.1007/s11947-010-0434-1. Key: citeulike:7908228.

Chaisawang, M. and Suphantharika, M. (2005). Effects of guar gum and xanthan gum additions on physical and rheological properties of cationic cassava starch. *Carbohydrate Polymers,* 61, 288-295.

Chen, M. H., Yeh, G. H. and Chiang, B. H. (1996).Antimicrobial and physicochemical properties of methylcellulose and chitosan films containing a preservative. *Journal of Food Processing and Preservation,* 20, 379-390.

Chien, P.J., Sheu, F. and Yang, F. (2007). Effects of edible chitosan coating on quality and shelf life of sliced mango fruit. *Journal of Food Engineering,* 78, 225-229.

Chirife, J., Ferro Fontán, C. and Benmergui, E.A. (1980). The prediction of water activity in aqueous solutions in connection with intermediate moisture foods. A_W prediction in aqueous non-electrolyte solutions. *Journal of Food Technology,* 15, 59-70.

Chung, D., Papadakis, S. E. and Yam, K. L. (2001). Release of propylparaben from a polymer coating into water and food simulating solvents for antimicrobial packaging applications. *Journal of Food Processing and Preservation,* 25, 71-87.

De Bernardi, L. A. (2002). http://www.alimentosargentinos.gov.ar/0-3/horta/Fecula Mandioca/Fecula_Mandioca_01.htm.

Delville, J., Joly, C., Dole, P. and Biliard, C. (2003). Influence of photocrosslinking on the retrogradation of wheat starch based films. *Carbohydrate Polymers,* 53, 373-381.

Devece, C., Rodriguez-Lopez, J., Fenoll L., Tudela J., Catala J., de Los Reyes E., and Garcia-Canovas, F. (1999). Enzyme inactivation analysis for industrial blanching applications: Comparison of microwave, conventional and combination heat treatments on mushroom polyphenoloxidase activity. *Journal of Agricultural and Food Chemistry*, 47(11), 4506-4511.

Devlieghere, F., Vermeulen, A. and Debevere, J. (2004). Chitosan: antimicrobial activity, interactions with food components and applicability as a coating on fruit and vegetables. *Food Microbiology*, 21, 703-714.

FAO (2004). Proceedings of the validation forum on the global cassava development strategy. Volume 6. Global cassava market study business opportunities for the use of cassava. *International fund for agricultural development*, Rome.

Flores, S.K., Famá, L., Rojas,A.M., Goyanes, S. y Gerschenson, L.N. (2007a). Physical properties of cassava-starch edible films. Influence of filmmaking and potassium sorbate. *Food Research International*, 40(2), 257-265.

Flores, S.K., Haedo, A.S., Campos, C. and Gerschenson, L.N. (2007b). Antimicrobial perfomance of potassium sorbate supported in cassava starch edible films. *European Food Research and Technology,* 225 (3-4), 375-384.

Flores, SK., Costa, D., Yamashita, F., Gerschenson, L.N. and Grossmann, M. V. (2010). Mixture design for evaluation of potassium sorbate and xanthan gum effect on properties of cassava starch films obtained by extrusion. *Materials Science and Engineering* C, 30, 196–202.

Franssen, L.R. and Krochta, J.M.(2003). Edible coatings containing natural antimicrobials for processed foods. In S. Soller (Ed.), *Naturals antimicrobials for the minimal processing of foods* (pp. 250-262). Boca Raton, FL: CRC Press.

Franssen, L. R., Rumsey, T.R. and Krochta, J.M. (2002). *Modeling of natamycin and potassium sorbate diffusion in whey protein isolate films for application to cheddar cheese. Poster* 28-5. Anaheim, California: Institute of Food Technologists Annual Meeting.

Gerschenson, L.N. and Campos, C.A. (1995). Sorbic Acid Stability during Processing and Storage of High Moisture Foods. In G. Barbosa-Cánovas and J. Welti-Chanes, *Food preservation by moisture control. Fundamentals and applications* (pp.761-790). Lancaster, PA: Technomic Publishing Co. Inc.

Greener-Donhowe, I. and Fennema, O. (1994) Edible Films and Coatings: Characteristics, Formation, Definitions, and Testing Methods. In J. M.

Krochta, E. A. Baldwin and M. O. Nisperos-Carriedo, *Edible coatings and films to improve food quality* (pp. 1-24). Lancaster, Pennsylvania: Technomic Publishing Co., Inc.

Guilbert, S. (1988). Use of superficial edible layer to protect intermediate moisture foods: application to the protection of tropical fruit dehydrated by osmosis. In C. C. Seow, *Food Preservation by Moisture Control* (pp. 199-219). Essex, England: Elsevier Applied Science Publishers Ltd.

Howard, L. R. (1996) Minimal processing and edible coating effects on composition and sensory quality of mini-peeled carrots. *Journal of Food Science*, 61, 643-651.

Jeon, Y.I., Kamil, J.Y.V.A. and Shahidi, F. (2002). Chitosan as an edible invisible film for quality preservation of berring and Atlantic cod. *Journal of Agricultural and Food Chemistry,* 50, 5167-5178.

Kester, J.J. and Fennema, O.R. (1986). Edible films and coatings: a review. *Food Technology,* December, 47-59.

Kim, S.J. and Ustunol, Z. (2001) Sensory attributes of whey protein isolate and candelilla wax emulsion edible films. *Journal of Food Science*, 66, 909-911.

Liu, Q. (2005) Understanding Starches and Their Role in Foods. In S.M. Cui, *Food Carbohydrates: Chemistry, Physical Properties and Applications* (pp. 309-355).Boca Raton, FL: CRC Press.

McHugh, T. H. and Senesi, E. (2000) Apple wraps: A novel method to improve the quality and extend the shelf life of fresh-cut apples. *Journal of Food Science*, 65(3), 480-485.

Mei, Y., Zhao, Y., Yang, J. and Furr, H.C. (2002). Using Edible Coating to Enhance Nutritional and Sensory Qualities of Baby Carrots. *Journal of Food Science,* 67(5), 1964–1968.

Ozdemir, M. and Floros, J.D. (2001) Analysis and modeling of potassium sorbate diffusion through edible whey protein films. *Journal of Food Engineering,* 47,149-155.

Sanjurjo, K., Flores, S.K., Gerschenson, L.N. and Jagus, R. (2006). Study of the perfomance of nisin supported in edible films. *Food Research International*, 2006, 39, 749-754.

Sebti, I., Blanc, D., Carnet-Ripoche, A., Saurel, R. and Coma, V. (2004). Experimental study and modeling of nisin diffusion in agarose gel. *Journal of Food Engineering*, 63, 185-190.

Suppakul, P., Miltz, J., Sonneveld, K. and Bigger, S.W. (2003). Active packaging technologies with an emphasis on antimicrobial packaging and its applications. *Journal of Food Science* 68(2), 408-420.

Thunwall, M., Kuthanová, V., Boldizar, A. and Rigdahl, M. (2008). Film blowing of thermoplastic starch. *Carbohydrate Polymers*, 71, 583-590.

Vasconez, M.B., Flores, S.K., Campos, C., Alvarado, J. and Gerschenson, L.N. (2009). Antimicrobial activity and physical properties of chitosan - cassava starch based edible films and coatings". *Food Research International*. 42 (7), 762-769.

Veiga-Santos, P., Oliveira, L.M., Cereda, M.P., Alves, A.J. and Scamparini, A.R.P. (2005). Mechanical properties, hydrophilicity and water activity of starch-gum films: effect of additives and deacetylated xanthan gum. *Food Hydrocolloids*, 19, 341-349.

Vodjani, F. and Torres, J. A. (1989a). Potassium sorbate permeability of polysaccharide films: chitosan, methylcellulose and hydroxypropyl methycellulose. *Journal of Food Processing and Engineering*, 58, 33-48.

Vodjani, F. and Torres, J. A. (1989b). Potassium sorbate permeability of edible celullose ether multi-layer films: effect of fatty acids. *Journal of Food Processing Preservation*, 13, 417-430.

Zobel, H. F. (1994). Starch granule structure. In R. J. Alexander and H. F. Zobel, *Developments in carbohydrate chemistry* (pp. 1-36). St Paul, Minnesota: The American Association of Cereal Chemists.

In: Cassava: Farming, Uses, and Economic Impact   ISBN:978-1-61209-655-1
Editor: Colleen M. Pace   © 2012 Nova Science Publishers, Inc.

Chapter 6

Innovations of Cassava Starch as Biodegradable Polymer Material

Lee Tin Sin[1,2], W. A. W. A. Rahman[1] and M. S. N. Salleh[1]

[1]Department of Polymer Engineering,
Faculty of Chemical Engineering,
Universiti Teknologi Malaysia,
81310 UTM Skudai, Johor, Malaysia
[2]Department of Chemical Engineering
Faculty of Engineering and Science,
Universiti Tunku Abdul Rahman
53300 Setapak, Kuala Lumpur

Abstract

Biodegradable polymers have gained great attention of researchers in decades ago. Biodegradable polymers are environment friendly products which are able to reduce environmental pollution problems. In recent decade, one of most important in the development of biodegradable polymer area is to produce cheap starch based biodegradable polymer. Native starch is suitable to produce biodegradable polymer material because it is available abundantly at low cost. Starch is harvested from varieties of crops such as corn, potato, sago, cassava, wheat and etc. Among the crops, cassava is most widely growth to produce sustainable and cheap source of starch globally. In this chapter, cassava starch is utilized to produce biodegradable polymer compound. Cassava starch

(CSS) is required to undergo series of processes in order to produce the CSS based polymer products by existing polymer injection moulding technology. CSS is blended with polyvinyl alcohol, glycerol, and processing aids to improve the processability by injection moulding. For instance, the polyvinyl alcohol is incorporated to enhance the physicomechanical properties of the CSS. Meanwhile, glycerol is added to CSS for lubrication and gelatinization purpose. All these ingredients need to be melt compounded by twin screw extruder into pellet form. The PVA-CSS polymer compound can be used to produce plastic articles by using injection moulding technique under optimum processing conditions. The amount of CSS in polymer compound influences the mechanical and thermal properties of the polymer compounds.

Keywords: Polyvinyl alcohol;Cassava Starch; Injection moulding; Mechanical properties; Thermal properties.

ABBREVIATION

AFNT	Approximated flat slab NTA
CSS	Cassava starch
DSC	Differential scanning calorimetry
NTA	Name tag shape article
PPV55	PPVA-starch blend containing 50 wt.% of CSS
PPVA	Glycerol plasticized poly(vinyl alcohol)
PVA	Poly(vinyl alcohol)
TGA	Thermogravimetry analysis

1. INTRODUCTION

Plastics especially polyethylene and polypropylene are being used throughout the world. Every year, over 125 million tons of plastics are tremendously consumed by human beings. Plastics are attractive materials because they can be used in wide range of applications. Although plastics give tremendous advantages to mankind, these polymers are traditionally designed resistance to microbial attack and biodegradation. Therefore, plastics are not environmental friendly and able to cause long term damage to the natural surrounding. In order to overcome this problem, the used of biodegradable natural polymer has been recognized as an alternative to replace the conventional plastics.

Starch can be used as a matrix to develop biodegradable polymers due to its fully biodegradable properties and low cost production. However, there are some drawbacks when using starch as a matrix, such as incompatibility with a hydrophobic synthetic polymer as well as having tendency to degrade during processing at high temperature. Different approaches have been adopted to utilize starch to combine with various types of synthetic thermoplastic polymers for the production of totally and partially biodegradable polymeric compound. In this case, the structure of the native starch should be modified prior blending with synthetic polymers. This is crucial because starch is a multi hydroxyl polymer. There are huge numbers of intermolecular and intramolecular hydrogen bonds in starch structure and thus native starch is not a true thermoplastic. But in the presence of plasticizer at high temperatures (90-180 °C) and shear effects, the starch is readily to melt and flow. The plasticized starch (also known as thermoplastic starch) is suitable to be used for injection, extrusion or blow molding material, similar to most conventional synthetic thermoplastic polymers [1,2]. Previous studies shows starch like cassava starch has been widely used in edible films and coatings to provide protection to food products [3,4]. Lawton [5] has prepared starch-PVA films with the addition of glycerol and poly (ethylene-co-acrylic) as processing aids to produce excellent biofilms.

Although the modified starch can be processed like thermoplastics, the physicomechanical properties of the thermoplastic starch remain weak and unfavourable. Blending of synthetic polymers is able to improve the properties of thermoplastic starch. Both polar and non-polar types of synthetic polymers have been used to blend with thermoplastic starch. However, blending of non-polar synthetic polymer (i.e. polyethylene and polypropylene) with starch has shown loss of properties [6]. In contrast, starch blending with polar polymer such as poly(vinyl alcohol) (PVA) is compatible. Inherently, PVA is also a biodegradable polymer. Therefore, PVA-starch blend is suitable to develop a formulation of biodegradable polymer compound without undergoing substantial modifications. The blending PVA and starch were reported having excellent mechanical properties and barrier behaviour [7]. Chiellini et al. [8] also reported, the ongoing investigation in laboratories on the formulation and applicability of the mixtures of PVA and bio-based "fillers" from low-value agro-industrial waste has the potential of attaining eco-compatible articles. This eco-composite has special mean to experience environmental degradation at the end of their service life that eventually can be programmed.

2. PVA-Cassava Starch Blend

Poly(vinyl alcohol) (PVA) and starch are known as polar substances. Blending of PVA and starch is compatible because both of them possess hydroxyl (-OH) functional groups. The presence of hydroxyl functional groups lead to formation of strong hydrogen bonding. The formation of hydrogen bonding between PVA and starch has been proven by the studies done by Sin et al. [9] and Sin et al. [10] which bring to the synergistic interaction of the both components. Besides that, Siddaramaiah et al. [7] also found that starch filled PVA was slightly higher in tensile strength with the incorporation of 10 wt.% starch into PVA has shown the tensile strength increased from 268 to 279 kg/cm^2 in line with the percentage elongation at break which increased from 200 to 230%.

The formulation of PVA and cassava starch (CSS) blends can be prepared by common compounding polymer method. Typically, CSS and PVA are prepared in conventional extrusion machinery with the incorporation of appropriate additives and plasticizers. The plasticizer such as glycerol helps PVA and CSS to achieve better macromolecular chains mobility and miscibility.

2.1. Modification and Formulation Development

Both starch and PVA possess hydroxyl groups (—OH), thus the formation of hydrogen bonds in the blend tend to promote localized stability with improved miscibility of starch and PVA. Originally, PVA has limited end uses despite its excellent chemical, mechanical and physical properties which is mostly applied as a solution in water. This limitation partly due to the fact that PVA in the unplasticized state has a high degree of crystallinity and shows little thermoplasticity close to the occurrence of decomposition at about 170 °C. The thermal degradation becomes pronounced at 200 °C which is still below its crystalline melting point at 215 °C [11]. Therefore, it is important to plastisize PVA prior blending with CSS. The plasticized PVA (PPVA) is expected to form thermoplastic PPVA. Different types of glycol plasticizers can be used such as glycerol, sorbitol, propylene glycol and etc. Commonly, glycerol is used because it is cheap and food safe. The amount of glycerol for plasticizing PVA is 30-40 phr (part per hundred resin) and should not be exceeding 40 phr to avoid glycerol migration out from the PPVA matrix. Thermoplastic PPVA offers significant improvement over neat PVA with better flexibility and flowability. PPVA is easily to blend with the CSS and able to process in high pressure processing machinery such as injection moulding. Besides that, addition of processing aids and thermal stabilizer are also assisting the molten blend to flow smoothly during extrusion despite having higher shear stress and friction during compounding. The thermal stabilizer such phosphoric acid (0.5-1 phr) as well as processing aid such as calcium stearate (2-3 phr) are cheap to be used to achieve the precribed processability during both compounding and injection moulding process.

2.2. Processing of PPVA-Starch Blends

The preparations of PPVA-starch blend involving two compounding stages. Firstly is to prepare the PPVA resin using a twin screws extruder. It is followed by second stage compounding process to melt blending PPVA, starch and glycerol to produce PPVA-starch resin. The glycerol is again added at 20 phr in the second compounding stage to help CSS easily disperse in the PPVA matrix. Prior to compounding in extruder, it is important to physically mix the ingredients in a high speed mixer. The intimate mixing is essential to ensure the additives typically glycerol diffuse thoroughly and homogenously mixed in the PVA matrix.

As described above, the raw materials are melt blended together in a process called compounding to produce either palletized PPVA or PPVA-starch resin. Compounding is a process of feeding and dispersing of fillers and additives in the molten polymer. The compounded PPVA-starch resin is then injection moulded into desired plastic articles. A twin screw extruder machinery as shown in Figure 1 is commonly use to mature and melted all the materials to form the formulated blend. In the preparation the PPVA-starch blends for injection moulding, the first step is to plasticize the PVA into the form of thermoplastic PVA which is known as plasticized PVA (PPVA) in pellet form. After that, the PPVA-starch resin is produced by compounding the pelletized PPVA, starch and glycerol in the twin screw extruder again. Eventually, the pelletized PVA-starch can be processed into desired end product by using injection moulding or compression moulding.

PVA-starch blend are very sensitive to moisture and water. The hydroxyl groups (—OH) in both PVA and starch structure tend to attract moisture from the surrounding. Therefore, a drying step is necessary for PPVA-starch resin prior to compounding in order to avoid air trap/bubble defects of the end products. It is often drying of the PPVA-starch resin at 80 °C for 24 h before undergoing injection moulding process.

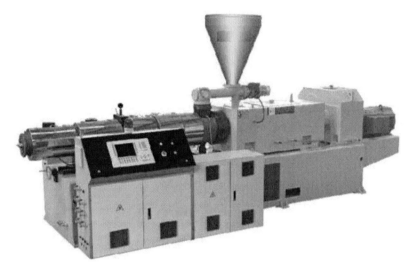

Figure 1. Industrial scale thermoplastic twin screw extruder.

2.2.1. Details of PPVA and PPVA-CSS Melt Blending/Extrusion Process

Extrusion is a process which all the raw materials are mixed and melted by using the extruder machine. There are two types of extruder which are single screw and twin screw extruder. Single screw extruder is not suitable for PPVA-CSS compounding due lack of shear effect and friction which is hardly allowed the ingredients to well blended and diffuse thoroughly. On the other hand, the twin screw extruder has one pair of screws rotate simultaneously. The high shear and friction of twin screw extruder enable the CSS and other ingredients effectively achieve a good dispersion throughout the blend matrix.

Normally an existing industrial scale twin screw extruder for melt compounding of thermoplastic is suitable to prepare both PPVA and PPVA-starch compounds. However, many of the twin screw extruders have relatively long screw design. In this case, a side feeder must be fully utilized. This is because the prolong residence time of molten PVA-starch compound in long screw extruder will tend to thermally degraded. Eventually, the PPVA-starch compound showed remarkable yellowish or brownish signing severe degradations have occurred. It is necessary to use the side feeder as a feeding zone in order to reduce the residence time with lower shear rate inside the extruder. Table 1 and 2 show the process parameters setting of Brabender Plasticorder PL2000 four zones laboratory scale co-rotating twin screw extruder to prepare PPVA and PPVA-starch resin.

Table 1. Temperature profile for compounding PPVA using Brabender Plasticorder PL2000 four zones laboratory scale co-rotating twin screw extruder

Extruder die	iv	iii	ii	i
$160\,^0C$	$160\,^0C$	$160\,^{\circ}C$	$160\,^{\circ}C$	$160\,^{\circ}C$

Screw speed: 250 rpm.

Table 2. Temperature profile for compounding PVA-starch using Brabender Plasticorder PL2000 four zones laboratory scale co-rotating twin screw extruder

Extruder die	iv	iii	ii	i
$140\,^{\circ}C$	$140\,^0C$	$140\,^{\circ}C$	$140\,^{\circ}C$	$140\,^{\circ}C$

Screw speed: 250 rpm.

2.1.2. Injection Moulding Process of PPVA-Starch

Basically the preparation of starch based compound by extrusion process is easier as compared to the production of the final product articles via injection moulding process. This is because the injection moulding of starch based compound possesses few difficulties such as high molten state viscosity and poor flow properties. Injection moulding is one of common conventional plastic processing equipment. Injection moulding able to produce discrete complex parts with variable cross sections cost effectively. Almost all thermoplastics and some thermosets can be injection moulded. The flexibility of the injection moulding technology is highly recognized globally [12]

Since thermal behaviour of starch is complex, thus CSS derived PPVA-starch has difficulty to be injection moulded as well. The injection moulding processing of starch compound is difficult to be controlled effectively due to its unique phase transitions, high viscosity, water evaporation, fast retrogradation, etc. [13]. The development of appropriate formulation and processing conditions for starch based compound are very important to ensure better processability in injection moulding machinery. In this chapter, the determination of optimum parameters for injection moulding PPVA-CSS blends was carried out using full factorial design approach. A full factorial design experiment helps to analyze of all possible combination factors and subsequently generate the reliable processing parameters for injection moulding of PPVA-CSS blends.

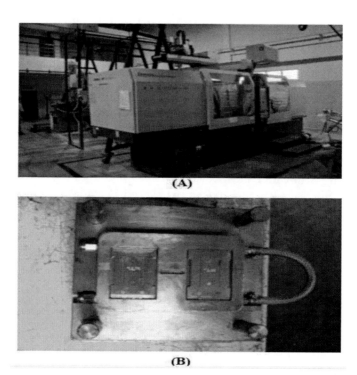

Figure 2. (A) Injection moulding machine Demag EL-EXISE 60-370 60 tons (B) Name tag shape article (NTA) mould.

In the PPVA-starch processability study conducted by the researchers in Department of Polymer Engineering, Universiti Teknologi Malaysia, four process factors were studied at 2-levels full factorial design. The four process factors were injection temperature, injection pressure, injection speed and packing pressure. The response of interest was volumetric shrinkage. A name tag shape article (NTA) mould was used as the reference product for process optimization. For determination of volumetric shrinkage, the average outer dimensions (length, wide, and thickness) of the injection moulded NTA were measured. The volume of the NTA mould dimensions were subtracted by the average volume of the PPVA-starch injection moulded NTA. According to 2-levels (high and low levels) full factorial design, there are 2^k of runs need to be conducted in order to produce comprehensive statistical analysis. k is the number of process factors [14]. Therefore, there were four process factors with a total of sixteen combinations which are possibly affecting the volumetric shrinkage. The influences of process parameters on the volumetric shrinkage was carried out in actual injection moulding machine Demag EL-EXISE 60-370 60 tons (Figure 2) with the general machine setting as shown in Table 3.

Table 3. Injection moulding machine setting

Mould temperature	40-45 °C
Holding time	2-3 s
Cooling time	12-15 s
Back pressure	5-6 bar
Screw speed	200-250 rpm
Screw retract	45-55 mm
Dosing	50-60 mm

2.1.3. Full Factorial Analysis Injection Moulding of PPVA-Starch Compound

Table 4 shows the settings of the process parameters and their respective levels for the injection moulding of PPVA-starch blend containing 50 wt. % of CSS (PPV55). Every set of combination was done in three replicates in order to obtain the standard deviation of the volumetric shrinkages. The standard deviation was used to study the variability effects of the process parameters.

Table 4. Process factors and level for injection moulding of PPV55

Sample	Process Factor	Low Level	High Level
	Injection pressure (bar)	80	100
	Injection temperature (°C)	200	220
PPV55	Injection speed (mm/s)	60	80
	Packing pressure (bar)	50	70

Initially, the outer dimensions (length, width, and thickness) were measured and recorded. Since NTA (Figure 3) is a very complex geometry and the measurement of the geometrical volume is very time consuming as well. In order to simplify the measurement of the complex geometry of NTA, the average of the measured length, width and thickness were multiplied to obtain the volume by approximating to a flat slab. Hence, the volumes of approximated flat slab NTA (AFNT) of PPV55 are recorded in Table 5, respectively. Similarly, the NTA outer mould dimension as shown in Figure 3 were also multiplied (90 × 59 × 4.50 = 23895 mm^3) to obtain the original volume. The differences of original and experiment dimensions of AFNT were further recorded in Table 6 as volumetric shrinkages of PPV55. Besides that, the standard deviations were calculated for variability analysis as well.

Generally, the volumes of AFNT for the above experimental design can be concluded very close to the original volume of the mould. However, different combinations of parameters have shown rather large variation among volumetric shrinkage and this indicated that interactions of the factors have occurred. An optimum set of process parameters can be obtained through a thorough statistical analysis.

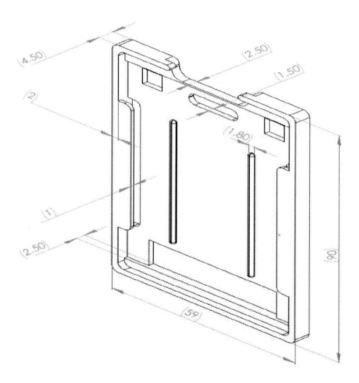

Figure 3. Drawing and dimensions of NTA.

Table 5. Measured volumes of NTA injection moulded of PPV55

No.	IT (°C)	IP (bar)	IS (mm/s)	PP (bar)	Volume (mm³)		
					1	2	3
1	200	80	60	60	23543	23105	23017
2	200	80	80	60	23643	23115	23127
3	200	80	60	80	23653	23313	23109
4	200	80	80	80	23644	23280	23317
5	200	100	60	60	23756	23817	23750
6	200	100	80	60	23760	23519	23417
7	200	100	60	80	23790	23777	23801
8	200	100	80	80	23799	23812	23883
9	220	80	60	60	22543	23002	22305
10	220	80	80	60	22641	23123	22339
11	220	80	60	80	22997	22899	23312
12	220	80	80	80	23012	23142	22937
13	220	100	60	60	22819	23133	23013
14	220	100	80	60	22919	22997	23001
15	220	100	60	80	23021	22819	23022
16	220	100	80	80	23114	22995	23101

IT = Injection Temperature IS = Injection Speed.
IP = Injection Pressure PP = Packing Pressure.

Table 6. Volumetric shrinkage of NTA injection moulded of PPV55

No.	IT (°C)	IP (bar)	IS (mm/s)	PP (bar)	Shrinkage (mm³)			SD
					1	2	3	
1	200	80	60	60	352	790	878	281.7398
2	200	80	80	60	252	780	768	301.4366
3	200	80	60	80	242	582	786	274.8187
4	200	80	80	80	251	615	578	200.3306
5	200	100	60	60	139	78	145	37.0720
6	200	100	80	60	135	376	478	176.1316
7	200	100	60	80	105	118	94	12.0139
8	200	100	80	80	96	83	12	45.2143
9	220	80	60	60	1352	893	1590	354.2913
10	220	80	80	60	1254	772	1556	395.4289
11	220	80	60	80	898	996	583	215.7923
12	220	80	80	80	883	753	958	103.7224
13	220	100	60	60	1076	762	882	158.4466
14	220	100	80	60	976	898	894	46.2313
15	220	100	60	80	874	1076	873	116.9145
16	220	100	80	80	781	900	794	65.2763

IT = Injection Temperature IS = Injection Speed SD = Standard Deviation.
IP = Injection Pressure PP = Packing Pressure.

Hence, the above results were imported into Minitab 15 for analysis. The design of experiment (DOE) analysis was carried out at significant level (α = 0.05). The objectives of current analysis are to find out the most significant process factors affecting the volumetric shrinkage and finally determine the setting of process parameters with the lowest volumetric shrinkage.

The normal probability plot of AFNT injection moulded by PPV55 shows that the response behaves in normal distribution (Figure 4A).

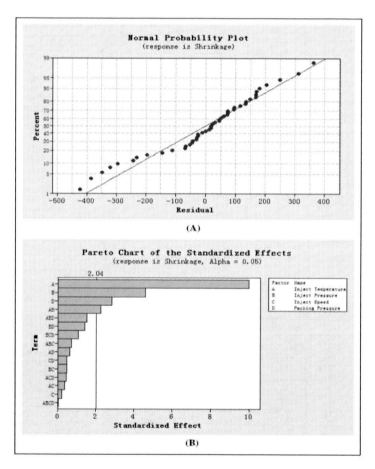

Figure 4. AFNT moulded by PPV55 (A) Pareto chart (B) Normal probability plot.

The injection temperature (A), injection pressure (B) and packing pressure (D) factors exhibited significant influence to the volumetric shrinkage of AFNT (Figure 4B). However, injection temperature and injection pressure factors have shown significant interactions with each others. The extent of this interaction is further analyzed through the interaction plot as shown in Figure 5. According to the interaction analysis, injection temperature and injection pressure should always set at 200 °C and 100 bar, respectively in order to achieve the lowest volumetric shrinkage of AFNT injection moulded by PPV55.

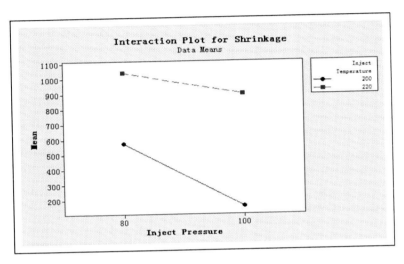

Figure 5. Interaction plot of AFNT moulded by PPV55.

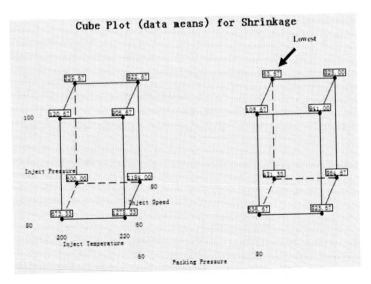

Figure 6. Volumetric shrinkage cube plot for AFNT of PPV55.

In addition to above analyses, cube plots of AFNT injection moulded by PPV55 is plotted in Figure 6. Cube plot is used to determine the best setting of the process factors to achieve the lowest volumetric shrinkage of AFNT. Figure 6 shows that the lowest volumetric shrinkage (63.67 mm^3) of AFNT moulded by PPV55 is setting at injection temperature 200 °C, injection pressure 100 bar, injection speed 80 mm/s, and packing pressure 80 bar. This setting agrees with the interaction plot in Figure 5 and further defines the injection speed and packing pressure as well.

In order to further confirm the results of the above analysis are correct, a verification experiment was conducted for PPV55. The results as recorded in Table 7 show that volumetric shrinkage of the AFNT agreed with the statistical analysis. These have proved that the optimization processing parameters for PPV55 to produce AFNT have been proven for its validity.

The second objective of the statistical analysis is to determine the process factors which affect the variability of the volumetric shrinkage of AFNT when injection moulded by PPV55. The standard deviation data as recorded in Table 6 are used in this analysis. However, the standard deviations were converted to logarithm so that the standard deviation will be closer to normal distribution condition [14] for Pareto chart. In this analysis, a large standard deviation of the particular setting is attributed to large differences of volumetric shrinkage among the NTA. This means that a high possibility of the products out of specification.

As shown in the plots in Figure 7, Injection pressure (B) process is the only factor has significant effect on the variation of volumetric shrinkage of AFNT when injection moulded by PPV55. None of the process factors have shown interactions with each other. This may be attributed to a higher composition of PVOH in PPV55 improve the stability of the polymer compound. The best setting to minimize variations of volumetric shrinkage is injection temperature 200 °C, injection pressure 100 bar, injection speed 60 mm/s, and packing pressure 80 bar. However, this setting does not agree has the minimum volumetric shrinkage of AFNT injection moulded by PPV55.

Table 7. Volumetric shrinkage of verification experiment

Filling Material	IT (°C)	IP (bar)	IS (mm/s)	PP (bar)	Shrinkage (mm^3)			Mean
					1	2	3	
PPV55	200	100	80	80	65	45	77	62.33

IT = Injection Temperature IS = Injection Speed SD = Standard Deviation.
IP = Injection Pressure PP = Packing Pressure.

At this point, selection the setting that produces minimum volumetric shrinkage is more preferable than low variability volumetric shrinkage setting because NTA is not marketed for high precision applications. Finally, the injection moulded NTA of PPV55 is shown in Figure 8.

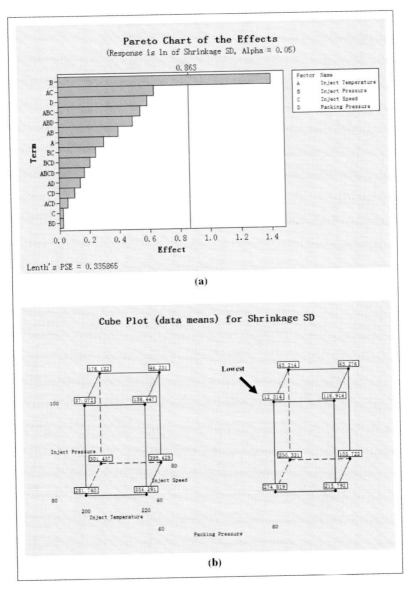

Figure 7. (a) Pareto chart (b) Cube plot of AFNT filled by PPV55.

Figure 8. Injection moulded NTA final product by PPV55.

2.3. Mechanical Properties of PPVA-Starch Blends

The mechanical properties of PPVA-starch blends such as tensile strength and elongation at break depend on the amount of starch and glycerol added into the formulated blend. Starch is added to reduce the amount of synthetic polymer in the composites for cost saving, while allowing greater speed and ease of biodegradation [15]. In a study conducted by Salleh et al. [16] the PPVA-starch blends were injection moulded to form tensile testing bars according to ASTM Standard [17]. The maximum content of starch in PPVA-starch blend that can be injection moulded to produce tensile bars was 80 wt. %. Increasing of starch content in the PPVA-starch blends have encountered processing problem such as short shot due to high melt viscosity lead to flow resistance when filling a narrow mould cavity.

2.3.1. Tensile Strength and Elongation at Break

The tensile test of various PPVA-starch blends was carried out to determine the mechanical performance as well as assessing the general relaxation behaviour of the materials external loading. Originally, PPVA is ductile, ability to yield and necking, strain hardening and eventually ruptures after subjected to a long elongation. The addition of granular CSS into PPVA matrix follows the general trend of filler effects on polymer performance. The tensile strength and elongation at break decreased gradually with increasing starch content. Without the addition starch, PPVA itself possesses the highest tensile strength at 13.085 MPa. This indicated that when PPVA continuous phase matrix is not disrupted, the matrix has higher flexibility and strength. Mao et al. [18] reported that the addition of starch has significantly disrupted the surface crack which led to the decrement in mechanical properties. Incorporation of 70 wt. % above of starch in the PPVA matrix was recorded the highest drop in tensile strength (5.29 MPa). However, a slight drop in tensile strength was observed after increasing the starch from 50 to 60 wt. % (7.9525 to 7.6339 MPa). This showed that an optimum starch content has been reached at 50-60 wt. % with good compatibility between PPVA and starch.

Apart from that, Young's modulus is determined in the tensile analysis as well. The increased in Young's modulus was observed from 37.889 MPa to 335.34 MPa when starch loadings were increased from 0% to 80 wt. %, respectively. The Young's modulus increased mainly due to the stiffening effect of the CSS granules. Starch is capable of withstanding higher stress at low strain as compared to PPVA. The rigid CSS granules have stiffened the PVA matrix. This is in agreement with Khalil et al. [19] and Cinelli et al. [15] which also reported similar observations in their works. It is known that starch is build up of D-glucose unit containing amylose and amylopectin [20]. These compositions are the major influences that contribute to the rigidity behaviour of CSS filler and becoming significant to induce higher Young's modulus.

Generally the tensile behaviour of PPVA-CSS blends exhibited brittle to ductile behaviours eventually failures when subjected to external straining forces. The PPVA showed that it could withstand over 500 N loads and extend up to 350 mm with promising ductile behaviour. However, blending of PPVA with 80 wt. % CSS noted a brittle behaviour indicated of brittle failure. It was reported the extension at break was only 29% of PPVA. Tensile curves of PPVA with 50-70 wt. % starch showed the trends towards brittleness failure. The addition of CSS caused the composite changing towards brittle failure with starch properties were predominating in the blends. Tensile curve behaviour also indicates that the incorporation of starch lowered the ability of blends to withstand the applied load. It can be proven by PPVA (without starch) and PPVA with 80 wt. % CSS which have maximum load to withstand were 500 N and 200 N, respectively. The starch granule blocked the PVA matrix from closed packing each other. Starch tends to promote amorphous effect in the blends, thus lowering the load required to break the matrix.

On the other hand, the elongation at break showed prominent decrement with increasing of CSS. It was found that the elongation at break decreased with higher starch content. Again starch played important role in affecting the blends matrix through disturbing the arrangement of continuous phase of PPVA. The reduction of tensile strength and elongation at break was due to the weak interfacial adhesion of the two components. This may indicate that the incorporation of CSS granules into the PPVA matrix introduces a new interfacial region that affects the stress transfer in the blends. Moreover, the addition of starch has reduced the elongation of the composite thereby decreased the flow of the continuous phase PVA [15]. The large starch particles could not elongate longer and could not flow along with PVA, thus starch can be pulled away from the matrix easily [15].

2.4. Thermal Stability of PPVA-Starch Blend

The thermal stability of PPVA-starch blends was investigated to determine the degradation point of the blends. Blending possesses high thermal resistance is favourable in order to withstand severe heating and shearing of injection moulding process.

2.4.1. Thermogravimetric Analysis (TGA) Analysis

Thermogravimetric analysis (TGA) was conducted to determine the thermal stability of PPVA-starch blends. The mass losses of the sample were recorded continuously while the temperature was increased at a constant rate [21]. At low temperature, the mass loss occurs when volatiles such as plasticizer and lubricants in the polymer blends were driven off. While at higher temperatures, degradation of the polymer occurs with the formation of volatile products due to chain scissioning or depolymerizations [22].

Thermal analyses of PPVA-starch blends at different starch composition were summarized in Table 8. With the incorporation of starch into the blends, the onset degradation temperature was shifted to higher temperature than PPVA itself. This is because PPVA-starch blend consists of starch component which is originally built up of cyclic hemiacetal [23].

Table 8. Percent (%) mass loss at onset degradation temperature

Blends	Mass Loss at Onset Degradation Temperature (%)	Onset Degradation Temperature (°C)
PPVA	77.15	279.41
20 wt.% PPVA -80 wt.% CSS	76.72	308.43
30 wt.% PPVA -70 wt.% CSS	65.47	307.71
40 wt.% PPVA -60 wt.% CSS	66.93	303.36
50 wt.% PPVA -50 wt.% CSS	72.37	301.01

The cyclic form of D-glucose unit arrangement has compact closed structure with shielding effect and thus higher energy is required to break the bonding [23].

2.4.2. Differential Scanning Calorimetry (DSC) Analysis

Differential scanning calorimetry (DSC) is one of the most widely used thermal analysis techniques to characterize polymeric material. Thermal events such as melting, recrystallization, and glass transitions can be well identified using DSC. Additionally, quantitative mixture analysis such as composition of polymer blends can be performed as well. The use of modulated DSC expands the capabilities of DSC and allows one to measure heat capacities and characterize reversible/non-reversible thermal transitions.

Thermograms of PVA-CSS blends obtained from DSC analysis are illustrated in Figure 9. Samples of PPVA, and PPVA blended with 50 wt. % and 60 wt. % of CSS have prominent endothermic peaks indicative of crystallinity in the PPVA-CSS blends. While blends consisted of 70 wt. % and 80 wt.% of CSS do not exhibit clear endothermic peaks which indicates the blends was actually in amorphous phase. This implies that as CSS increases in PPVA blends, the irregularity structure of starch has disrupted the structural crystalline arrangement of PVA and resulted the PVA-CSS loss of the strong direct interactions. Besides that, Table 9 showed that the incorporation of CSS into the PPVA matrix has led to the decrement in the onset, end-point and melting temperatures (T_m) of the blends. T_m is related to kinetic energy required to break the chain out of the rigid structure while the enthalpy of fusion (area under curve) can determine the extent of crystallization interaction in the blends [23]. Higher amount CSS has disrupted the genuine interaction of neat PVA, subsequently lower kinetic energy is required to break the chain out of the rigid structure. Nevertheless, the amount of plasticizer, i.e. glycerol also plays important in affecting the thermal properties as well. Furthermore, DSC analysis in Table 10 also recorded that the enthalpy of fusion (ΔH_m) for 50 to 80 wt.% of CSS in the blends are very much lower that the PPVA. This reflected that the lack of compatibility of PVA and starch structure has reduced the total energy requirement to achive molten state of the blends. Although addition of starch has disrupted the structure of PVA-CSS blends, on the other side, PVA can be considered as the effective binding agent for starch. This is because the addition of PVA has led to the strengthen the weaken structure of starch. PVA has acted as the binding agent of the starch granules which is shown as the apprearance of ΔH_m of 1.99 J/g with addition 20 wt. % of PPVA. Therefore, the blending of PVA and CSS is still considered a positive approach to provide better properties of the blends.

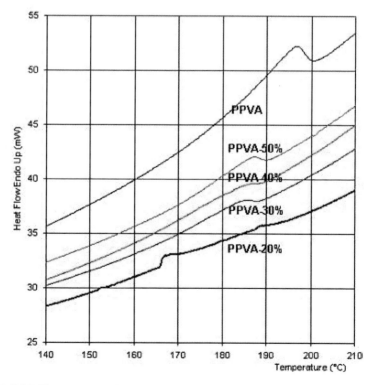

Figure 9. DSC Thermogram of PPVA-CSS blends.

Table 9. Onset and end-point melting temperature, melting temperature (T_m) and enthalpy of melting (ΔH_m)

Sample	Onset (°C)	End-point (°C)	ΔH_m (J/g)	T_m (°C)
PPVA	183.41	200.00	14.92	196.03
PPVA-50%	173.89	189.97	6.59	186.37
PPVA-40%	175.70	192.20	4.09	184.50
PPVA-30%	170.41	189.48	3.32	181.20
PPVA-20%	165.78	177.83	1.99	166.83

CONCLUSION

The utilization of cassava starch (CSS) as the biodegradable polymer material is an innovation. CSS is blended with poly(vinyl alcohol) (PVA) in order to improve the processability, mechanical, and thermal properties of the blends.

Prior to produce the biodegradable plastic article, all ingredients (CSS, PVA, glycerol, and additives) need to be melt blended in a twin screw extruder to produce PVA-CSS resin. PVA-CSS resin can be well processed by injection moulding technology with the selection of appropriate processing parameters. Generally, blending high amount (<50 wt. %) of CSS into PVA is not favourable due to loss of mechanical and thermal properties. Generally, continual improvements of the starch based biodegradable polymer compound are crucial to further develop products with excellent properties. Such attempts by the researchers can widen the applications of cassava starch biodegradable polymer materials in the future.

ACKNOWLEDGMENTS

This project is financially supported by Ministry of Science, Technology and Innovations (MOSTI) of The Federal Government of Malaysia-Putrajaya under E-Science Fund 03-01-06-SF0468 and the National Science Fellowship 1/2008.

REFERENCES

[1]   Curvelo, A. A. S.; Carvalho, A. J. F.; Agnelli, J. A. M. *Carbohydrate Polymers* 2001, 45, 183-188.

[2]   Ma, X. F.; Yu, J. G.; Wang, N. *Carbohydrate Polymers* 2007, 60, 111-116.

[3]   Flores, S.; Fama, L.; Rofas, A. M.; Goyanes, S.; Gerschenson, L. *Food Research International* 2005, 40, 257-265.

[4]   Fama, L., Rojas, A. M., Goyanes, S., and Gerchenson, L. LWT- *Food Science and Technology* 2005, 38, 631-630.

[5]   Lawton, J. W. *Carbohydrate Polymers* 1996, 29, 203-208.

[6]   Rahmat, A. R., W. A. W. A. Rahman, Sin, L. T., and Yussuf, A. A. *Materials Science and Engineering*: C 2009, 29, 2370-2377.

[7]   Siddaramaiah, Raj, B., and Somashekar, R. *Journal of Applied Polymer Science* 2004, 91, 630-635.

[8]   Chiellini, E., Cinelli, P., Chiellini, F., and Iman, S. H. *Macromolecular Bioscience* 2004, 4, 218-231.

[9]   Sin, L. T.; Rahman, W. A. W. A.; Rahmat, A. R.; Khan, M. I. *Carbohydrate Polymers* 2010, 79, 224-226.

[10] Sin, L. T.; Rahman, W. A. W. A.; Rahmat, A. R.; Samad, A. A. *Polymer* 2010, 51, 1206-1211.

[11] Famili, A; Nageroni, J. F.; Finn, L. M . 1994. U.S. Patent 5,362,778, Washington DC: U.S. Patent and Trademark Office.

[12] Strong, B. A. Plastics Materials and Processing; 3rd edition; Prentice Hall: *Upper Saddle, NJ*, 2006, pp 431-495.

[13] Liu, J.; Xie, F.; Yu, L.; Chen, J.; Li, L. *Progress in Polymer Science* 2009, 34, 1348-1368.

[14] Antony, J. *Design of Experiments for Engineers and Scientists*. Elsevier: Netherlands, 2003.

[15] Cinelli, P.; Chiellini, E.; Lawton, J. W; Iman, S. H. *Journal of Polymer Research* 2006. 13, 107-113.

[16] Salleh, M. S. N.; Rahman, W. A. W. A.; Sin, L. T. Tensile behaviour and thermal analysis of biodegradable injection grade tapioca starch filled plasticized poly(vinyl alcohol). Environmental Science and Technology Conference (ESTEC2009), Kuala Terengganu Malaysia, 7-8 December 2009, 485-493.

[17] ASTM Standard. D638 Standard Test Method for Tensile Properties of Plastics. ASTM International: West Conshohocken, PA, 2010. DOI: 10.1520/D0638-10.

[18] Mao, L.; Imam, S.; Gordon, S.; Cinelli, P.; Chiellini, E. *Journal of Polymers and the Environment* 2000, 8, 205-212.

[19] Khalil, A. H. P. S.; Rozman, H. D. *Polymeric Plastic Technology Engineering* 2000, 39, 757-781.

[20] Wade, L. G. Organic Chemistry; 4th edition; Prentice Hall: NJ: Prentice Hall, 1999.

[21] Campbell, D.; White, J. R. Polymer Characterization Physical Techniques. *Chapman and Hall*: New York, 1989.

[22] Price, D. M.; Hourston, D.J.; Dumont, F. (2000). In Thermogravimetry of Polymers; Meyers, R. A.; Eds.; Encyclopaedia of Analytical Chemistry; John Wiley: Chichester, 2000, 8094-8105.

[23] Rahman, W. A. W. A.; Sin, L. T.; Rahmat, A. R.; Samad, A. A. *Carbohydrate Polymers* 2010, 81, 805-811.

Reviewed by
Dr. Mohd Halim Shah Ismail, CEng. MIChemE,
Department of Chemical and Environmental Engineering
Faculty of Engineering,
Universiti Putra Malaysia,
Malaysia

In: Cassava: Farming, Uses, and Economic Impact   ISBN:978-1-61209-655-1
Editor: Colleen M. Pace   © 2012 Nova Science Publishers, Inc.

Chapter 7

CASSAVA:
A MULTI-PURPOSE CROP FOR THE FUTURE

Anna Westerbergh[1], Jiaming Zhang[2] and Chuanxin Sun[1][1]

[1]Department of Plant Biology and Forest Genetics, Uppsala BioCenter,
Swedish University of Agricultural Sciences,
Uppsala, Sweden
[2]Institute of Tropical Bioscience and Biotechnology,
Chinese Academy of Tropical Agricultural Sciences,
Hainan Province, China

ABSTRACT

Cassava, ranking presently as the world′s fifth largest crop in starch production, is shedding light on its importance in the global agricultural economy due to its various biological characteristics. It is a perennial crop, easy and economical in cultivation. It requires little or no fertilizers, beneficial for environments. Cassava is economical in land usage. It can utilize low quality land such as semi-dry and mountainous land. Importantly, it can produce large tuberous roots that result in a high

---

[1] Author for correspondence: Dr. Chuanxin Sun. Department of Plant Biology and Forest Genetics, Uppsala BioCenter, Swedish University of Agricultural Sciences, P.O. Box 7080, SE-750 07 Uppsala, Sweden. Email: chuanxin.sun@vbsg.slu.se. Phone (work): +46-(0)18-673252. Phone (cell phone): +46-(0)736221540. Fax: +46-(0)-18-673389.

starch yield. Moreover, its above-ground biomass can be employed as industrial feedstocks. This review will, from an economic point of view, elucidate the importance and possible contribution of cassava to the world´s future agricultural economy. By analyzing the use-value, economic benefits, threats to cassava farming, directions of traditional and molecular breeding, possible added-value to the future cassava and the World production in the past twenty years, the possibility of whether cassava can be used as a multi-purpose crop to meet the requirements for future plants is discussed.

THE ORIGIN AND DOMESTICATION OF CASSAVA

Cassava is a perennial root crop in the tropics belonging to the genus *Manihot* in the family Euphorbiaceae (Rogers and Appan 1973). Cassava is the only cultivated taxon among the nearly 100 *Manihot* species. The wild *Manihot* species are all endemic to the neotropics in South and Central America. Many of them have evolved in the semi-arid and savannah ('cerrado') biomes and are therefore adapted to dry ecosystems with seasonal drought and frequent fires. Other *Manihot* species grow in Amazon forest areas with no dry season. Cassava originates from the southwestern Amazon Basin of Brazil and has been domesticated perhaps as long as 8,000-10,000 years ago from its wild ancestor *M. esculenta* ssp. *flabellifolia* (Olsen and Schaal 1999, 2001).

Cassava was brought from South America by the Portuguese traders. It was first introduced at the West African coast in the late 16th century and then spread along the coast and to the inland through trade roads (Jones 1959, Carter et al. 1992). About two centuries later, the crop was introduced to the East African coast. It spread to the west and east and reached South East Asia at the end of the 18th century.

Selection during domestication has resulted in many morphological, physiological and biochemical differences between cassava and its wild relatives. Some traits such as increased size of tuberous roots and vegetative propagation through stem cuttings are results of human selection. Other traits such as higher protein content in the roots, lower post-harvest physiological deterioration (PPD) and resistance to pests and diseases were lost during domestication and are only preserved in the wild relatives. The genetic and phenotypic changes as a result of domestication are investigated by the first author and her research group.

THE USE-VALUE OF CASSAVA

One of the important reasons why cassava has been able to spread throughout the tropical regions is that cassava has a very good use-value. It can produce large starchy tuberous roots with many applications. It also produces a large amount of stem and leaf material that can be used as a vegetable, feedstocks and fodder.

Tuberous roots: Cassava is mainly grown for its starchy tuberous roots. A majority of the cassava farmers are smallholders living in poor rural communities in Africa, Asia and Latin America. Most of the cassava root production is for local consumption and the starchy tuberous roots are the third most important source of calories in the tropics consumed by more than 600 million people on a daily basis. Besides use of tuberous roots for food, they are also used as a source of starch and farmers usually sell the raw material to starch factories where it is processed. Cassava processing is a relatively new business and export volumes are still low in many cassava growing countries. The tuberous roots are also used for animal feed and sold to animal feed factories. Here, the roots are processed into pellets and chips. The interest of using cassava roots for production of ethanol as biofuel is increasing and will further enhance livelihood opportunities for the low-income rural farmers (Kim et al. 2008). The waste root material from ethanol production may also serve as an alternative source for animal feed.

Leaves and stems: Young leaves of cassava are, in some areas, especially in sub-Saharan Africa, used for human consumption as a vegetable (Lancaster and Brooks 1983). The leaves are rich in proteins, vitamins and minerals and serve as an important supplement of the predominant starchy diet in the cassava cultivating areas (Nweke et al. 2002). Besides tuberous roots, leaves and stems are also used as animal feed. The nutritional value is similar to alfalfa and cassava is often referred to "tropical alfalfa". The leaves and stems can be dried, fermented for silage production and processed into protein-rich pellets in animal feed factories. The stems also have a regenerative function, while the tuberous roots are modified lateral roots, enlarged to function as a storage organ. The farmers divide the middle part of the stem into two-node long cuttings. These cuttings are planted directly in the soil and at the leaf nodes and the cut-surface adventitious roots are formed for successful plant establishment.

A cassava plant usually produces ca. 1.0 kg dry weight (depending on variety) at a time of storage root maturation (at a time of harvesting) which is about 12-14 months after planting (El-Sharkblano 2004). Of the total dry weight of a cassava plant, roots, stems and leaves usually make up 45%, 35% and 20%, respectively (Figure 1). Commercial prices for the different tissues vary a lot on the current markets, mainly depending on regions where cassava is produced. In general, however, the price ratio of fresh tissues is 1:1:3 for stems, leaves and roots, respectively (Figure 1; Ebukiba 2010 and http//www.alibaba.com). Although the roots produce most of the dry weight (starch) with a good price, the leaves may have a good economic potential, particularly for those varieties used as fodder. They have a high fresh weight and can contain up to 20% protein content in some varieties (Ravindran 1993). A high fresh weight of leaves with a sustainable nutrition value may bring a tremendously economic benefit from the tissue, even as high as the tuberous roots, when they are used as a green vegetable for a healthy food and as fodder for livestock.

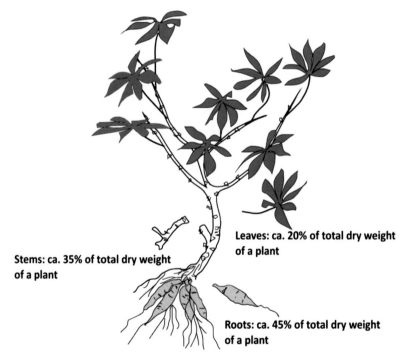

Leaves: ca. 20% of total dry weight of a plant

Stems: ca. 35% of total dry weight of a plant

Roots: ca. 45% of total dry weight of a plant

Price ratio for fresh tissues (stem:leaf:root): ca. 1:0-1:3

Figure 1. Schematic representation on a typical cassava plant that produces different portions of tissues with different prices. On the current markets, the fresh weight price ratio is around 1:1:3 for stems, leaves and roots. The price ratio 0 for leaves means that fresh leaves of some varieties are not in use.

CYANOGENIC GLYCOSIDES IN LEAVES AND ROOTS

A potential problem of cassava consumption is the content of cyanogenic glycosides in leaves, stems and tuberous roots. Cyanogenic glycosides are synthesized in the leaves and translocated to the stems and roots (Koch et al 1992, Siritunga and Sayre 2003, Jørgensen et al. 2005). Upon tissue disruption, these cyanogenic glucosides are degraded and cyanide (HCN) is released (White et al. 1994). High glycoside levels in the roots are associated with bitterness and these varieties are called bitter, whereas varieties low in cyanogenic glycosides are called cool or sweet. The small-scale farmers are able to distinguish between the two type of varieties and predicting the relative content of cyanogenic glycosides in different roots by tasting the tuberous roots (Chiwona-Karltun 2004). Leaves and bitter tasting roots need to be processed in order to reduce the content of HCN, while roots of sweet cassava are considered safe to eat without processing. Traditional processing of cassava roots vary between regions but typically involves different combinations of a number of techniques such as grating, soaking, fermenting, sun-drying and roasting (Essers et al. 1995, Padmaja 1995). The HCN content in leaves is greatly reduced by sun-drying and boiling or by ensiling. Eating poorly processed tuberous roots can cause severe health disorders such as paralysis (Konzo disease) and neurological disorders (Tylleskär 1992). The proportion of sweet and bitter varieties in farmers' fields depends on their preferences and cultures (e.g. Elias et al. 2001, Balyejusa Kizito et al. 2007a). Farmers emphasize cooking and storage qualities and starch characteristics. In some areas where cassava is the main staple crop, farmers preferentially grow bitter varieties even though sweet varieties are available. Cyanogenic glycosides are proposed to act as a protection against thefts, herbivory by insects and mammals and attacks by pathogens (Chiwona-Karltun et al. 1998, Bellotti and Riis 1994, Riis et al. 2003).

BREEDING FOR BITTER AND SWEET CASSAVA

Even though cyanogen-free cassava represents a safer food product, it may not be accepted among farmers in all areas. To meet diverse needs, future breeding programs must consider using both bitter and sweet varieties as breeding material. In cases where varieties with different content of cyanogenic glycosides are crossed in a breeding program, it is of major importance to have knowledge about the genetic basis of "bitterness" vs "coolness" in cassava roots. By using quantitative trait loci (QTL), mapping accompanied with fine-scale mapping or other mapping methods, molecular markers associated with genes controlling this trait may be identified and applied in marker-assisted breeding. Even though the presence of low or high levels of cyanogenic glucosides in cassava is a major criterion for the adoption or rejection of a variety by farmers, it is also among the least understood agronomic traits in cassava. Studies have been conducted on the biochemical pathway (e.g. Du et al, 1995, Siritunga et al. 2004) but a lot remains to be understood on the genetic control of cyanogenic glucosides in cassava. To our knowledge, Balyejusa Kizito et al. (2007b) have taken the first step towards identifying QTL for cyanogenic glycosides in cassava roots.

An Under-Exploited Crop

Despite its importance in the tropics, we consider cassava an under-exploited crop for rural economic development, food security and poverty reduction. With new markets for food, starch, feed and ethanol production, cassava may increasingly be perceived as a major cash crop. This will increase the opportunity for small-scale farmers to earn money through commercialization of cassava and not totally rely on subsistence agriculture for their survival.

The major challenge will be to develop the infrastructure including acquisition, transportation of harvested cassava materials from the small farmers. The current infrastructure for processing of cassava materials are not well-shaped in most of the cassava farming regions.

Due to climate change, cassava may enhance its role as a food security crop. An estimated temperature increase of three degrees over the next century will lead to an extensive global climate change. This will change the field conditions to which cassava and other crops are adapted and thereby the farming activities. According to modeling of crop suitability on agricultural soils of Africa by Andy Jarvis and Julian Ramirez at the International Center for Tropical Agriculture (CIAT), Cali, Colombia for the use of cassava may become an important alternative to other crops, such as maize in the future due to the hardiness of the cassava to endure high temperatures (http:// www. slideshare.net/ciatdapa/climate-change-and-the-outlook-for-cassavanov-2010). Cassava also harbors other qualities which makes it an attractive crop for poor resource farmers. It is tolerant to drought and low pH and gives reasonable yields in poor soils. Cassava cultivation does not require high amount of labor, fertilizers or other economic inputs compared to many temperate crops. Unlike sexually reproduced seed crops, the preferred cassava genotypes are maintained between growing seasons through clonal propagation and no planting material needs to be purchased by the farmers. Instead, the farmers use stem cuttings from plants in their own fields or cuttings obtained from other farmers. Even though cassava is mainly vegetatively propagated, cassava shows high genetic diversity within and between farmers' varieties (Elias et al. 2001, Fregene et al 2003, Balyejusa Kizito et al. 2005, 2007a) probably as a result of local breeding by small-scale farmer management with minor influence from breeding programs. The genetic diversity makes cassava less vulnerable to pest and diseases as well as environmental changes. The farmers' fields serve as an *in situ* gene bank for breeders and more defined strategies for on-farm conservation need to be developed.

To enhance the role of cassava as a multi-purpose crop and to ensure that cassava will be an important source of income for rural farmers, the researchers and breeders have to put further emphasis on improved root quality, higher root yield, resistance to pest and diseases and increase the added-value for the future cassava (see sections below). In addition, to reduce malnutrition further, emphasis has to be towards improving the content of proteins, minerals and vitamins in cassava. The two international biofortification programs, BioCassava Plus and HarvestPlus, aims at enhancing the bioavailable levels of protein, zinc, iron, vitamin A and vitamin E in cassava roots.

THREATS TO CASSAVA

While cassava's importance is increasing in the tropics, it suffers from a plethora of post-harvest physiological deterioration of tuberous roots (PPD, see sections below), pests and diseases such as mealybugs, cassava bacterial blight and cassava mosaic disease that threaten production and reduce profitability. It has been estimated that cassava farmers, typically resource-poor farmers, lose 48 million tons of fresh root, some 30% of total world production, valued at US$1.4 billion every year to pests, diseases, and PPD (FAOSTAT 2004). This will be a threat to the small cassava farmer and to a sustainable development of rural communities dependent on agriculture for their livelihood.

THE BUDGET FOR CASSAVA FARMING

A well-done calculation on cassava farming in Nigeria has been published recently (Ebukiba 2010). The author came up with a profit rate of 190% or a benefit cost ratio 1.90:1.00 in the country. It means that if farmers put in a single US dollar for cassava farming, they will win 90 cents. To elucidate, if the economic benefits for cassava farming is just country-specific, here we do another careful calculation on the budget of cassava farming in China. Interestingly, we came to a very similar profit rate, i.e. 183%, or 1.83:1.00. This benefit is much higher than farming of rice, maize and sugarcane with profit rates of 171.0%, 128.0% and 151% in China, respectively (Tables 1, 2 and 3).

Input for Cultivating Cassava in China

The input for cultivating cassava includes planting materials, fertilizer, pesticides, land preparation and labor costs. Usually, 1.5 tons of stem cuttings are used in each hectare. The average cost of the stem cuttings is around 1200 CNY per ton. The total input in a typical plantation in Danzhou, Hainan, China was around 8700 CNY ha^{-1} in 2007, without counting land preparation costs. For details, see Table 1.

Table 1. Major input for growing cassava on a typical land in Danzhou, Hainan, China in 2007 (Li et al. 2008)

	Costs (CNY ha^{-1})
stem cuttings	1800
fertilizers	1500
pesticides	450
labor costs	4950
in total	8700

Output of Cassava in China

The production of cassava depends much on varieties, land quality and management. Table 2 shows the average yields and starch contents of varieties used in China.

The price of cassava fresh roots was ca. 360 CNY per ton in 2006, and it rose to 500 CNY per ton in 2010 due to the production of bioethanol from cassava, which put increased profit to the cultivation of cassava and made cassava as one of the most profitable crops in Hainan, China (Table 3).

Table 2. Yield of main varieties used in China (Huang et al. 2007)

Variety	Average yield	Highest yield	Starch content
Huanan 205	30-45	75	28-30
GR891	30-45	67	30-33
Huanan 124	30-45	75	24-27
Nanzhi 199	30-45	75	28-32
Huanan 5	30-45		30-35
Huanan 6068	15-23	45	30-35
Huanan201	23-38		25-28

FW, t/hm^2: fresh weight of roots, ton per hundred square meters.

Table 3. Economic comparison of the main crops in Hainan, China

Crops	Yield (Ton ha^{-1})	Average price (CNY kg^{-1})	Output (CNY, total)	Input (CNY)				Profit	Profit rate (%)
				Total	Materials	Labor	Land		
rice	6	2.4	14400	8400	4350	2250	1800	6000	171
maize	4.8	1.8	8640	6750	3300	2400	1050	1890	128
sugarcane	60	0.325	19500	12950	8800	2650	1500	6550	151
cassava	28	0.5	14000	7650	3750	3000	900	6350	183

Data from the Hainan Provincial Department of Agriculture, China.

ADDED-VALUE FOR THE FUTURE CASSAVA AS A MULTI-PURPOSE CROP

Modification of Cassava Starch to Improve Cassava Food Quality

Improvement of food quality by modifying starch structures has been carried out intensively in breeding history of many crops including cassava. The goal can be achieved by both traditional and molecular breeding. For cassava traditional breeding, we have crossed cassava with its wild ancestor *Manihot esculenta* ssp. *flabellifolia*. The F_2 progenies (second generation) showed an enormous difference in many phenotypic traits including number, weight and shapes of tuberous roots as well as starch content (Castelblanco et al. 2008). The F_2 genotypes may also create a gene pool where we can seek different types of starch quality. For molecular breeding, some recent studies have shown that cassava is transformable (Taylor et al. 2004) and its starch yield and quality can be modified by gene technology (Ihemere et al. 2006). We have previously cloned the genes for starch branching enzymes and debranching enzymes that are important players in determining amylopectin structures (Baguma et al. 2003, Beyene et al. 2010). By using a similar strategy to what has been reported in potato (Schwall et al. 2000), we may be able to create a cassava with a high content of amylose that is in a high demand for industrial feedstocks. We may also generate a truly sweet cassava by knocking down expression of the starch debranching enzyme. The debranching enzyme-deficient cassava may accumulate sucrose that gives a true sweet taste as reported for "sugary" mutants in many other crops (Burton et al. 2002).

Increases of Protein Contents in Storage Roots

Protein root content is very low in cassava cultivars compared to the wild ancestor (*M. esculenta* spp. *flabellifolia*) as a result of domestication (see the first section). This has brought a nutrition problem in the countries where cassava is used as a staple food, particularly in Africa. One quick way to increase the protein content is to introduce a storage protein into the roots of a transformable variety. The trait can be further introduced to any elite variety by traditional breeding. An example of protein transformation is by Zhang et al. (2003) who have successfully introduced some heterologous storage proteins to cassava tuberous roots. However, two concerns might be worthy of taking into consideration. Any heterologous proteins introduced by molecular breeding should not give any drawbacks for the future cassava such as any allergic threats to human health. Another important concern is that high protein content requires a good nitrogen source which is usually provided by fertilizers. Cassava is, in many cases, cultivated in soils poor in nutrients. Getting enough nitrogen for the production of high content proteins will be a challenging issue in cassava farming. One very speculative suggestion is a possibility of breeding cassava to a crop with ability to attract bacterial symbiosis for nitrogen fixation. A practical way may be to intercrop cassava with legumes.

Increase of Post-Harvesting Saccharification of Cassava Starch

Saccharification of cassava starch seems to be one of the bottlenecks in the industry of bioethanol production. We have previously proposed introduction of more amylopectin-bound phosphate groups in the cassava starch by over-expression of a heterologous glucan-water dikinase (GWD; Jansson et al. 2009). One could also introduce an inducible and efficient α-amylase to the tubers. The enzyme can help starch degradation when induced. In addition, any changes of amylose and amylopectin ratio and chain-length of amylopectin branches can also change water solubility of starch. Highly soluble starch can be obtained by changing activities of starch ground-bound synthase I (GBSSI) or branching and debranching enzymes as described in our previous publications (Baguma et al. 2003, Beyene et al. 2010).

Starch to Oil

Although using sucrose, starch and lingo-cellulose from higher plants to produce bioethanol is applicable, production of bioethanol would not be as efficient and economic as plant-derived oils (for a review, see Ohlrogge et al. 2009). The main reason is that energy loses during fermentation from starch to ethanol are large (Jansson et al. 2009). Thus, breeding of some high-yielding starch crops such as cassava to oil crops may bring considerably high profits.

Carbon metabolism in the cassava roots is a complex issue. Metabolic reactions involving hexoses take place in the cytosol and the amyloplasts of these cells. Most of the reactions in both compartments are characterized by an almost redundant set of enzymes catalyzing both anabolic and catabolic reactions. The metabolite pools of the cytosol and the amyloplast are efficiently connected by transporters for triose phosphate, hexose phosphate, pentose phosphate, and ADP-Glc, which enable metabolite flux and maintain the phosphorus balance in the different compartments of the cell. The incoming sucrose in the roots is the carbon source for a variety of metabolic pathways (Figure 2) that result in different end products i.e. starch and cellulose. In cassava, the route to storage fat (oil bodies) is very minute compared to starch synthesis. However in another root crop, yellow nutsedge

(*Cyperus esculentus* var. *sativus,* Turesson et al. 2010), significant amounts of oil (up to 24%) together with starch (32%) can be found in its roots. Any biochemical data achieved from yellow nutsedge can be applied to cassava for starch-to-oil research.

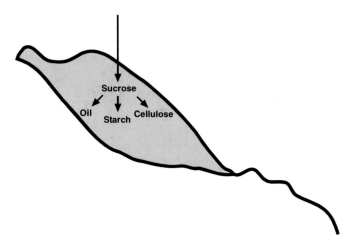

Figure 2. A cassava tuberous root where sucrose is uploaded and serves as carbon sources for different metabolic pathways.

Increase of a Shelf Life for Storage Roots

Another serious bottleneck in cassava production and commercialization for the industry of bioethanol production is the short shelf life for tuberous roots. The short shelf life (2-3 days) directly affects collection and transportation of cassava from small farmers to the market, the starch factories and the bioethanol power plants. Interestingly, Morante et al. (2010) has recently found that a few root sources with a much longer shelf life or with high tolerance against post-harvest physiological deterioration (PPD) have a significant amount of antioxidant of carotenoids.

The result suggested that the genes involved in carotene synthesis may be involved in determing the shelf life. They also found that the phenotypic trait of a long shelf life was also somehow associated with a wax gene (GBSS). Therefore, their finding and any further characterization of the genes or loci responsible for the shelf life will be a large contribution to cassava applications for both food and non-food industries.

Crossing the Latitude Limitation for Cassava Farming

Cassava is a tropical crop and can be more or less only grown in the tropical regions between 30°N and 30°S. One obvious challenge can be to "move" cassava farming up or down to a higher degree of latitude regions such as the regions of horse latitudes. This approach requires cold-tolerant varieties. A high mountain variety may clarify the climate of a low temperature. Introduction of some cold-tolerant genes may be another strategy (Wang et al. 2008).

Building up a Good Infrastructure for Cassava Post-Harvesting Processes on Both Food and Non-Food Applications

The high economic benefits may attract investors in the future to build up a better infrastructure for processing of cassava materials. A good infrastructure is very important for cassava farming in the future. In fact, the currently bad post-harvesting infrastructure limits scale up of cassava farming and is probably the major reason why cassava is cultivated mostly by small-scale farmers. The future infrastructure will include at least acquisition, transportation and processing for food and non-food applications, such as bioenergy, functional food, pharmaceutical materials, biodegradable plastics and many other industrial feedstocks.

CAN CASSAVA WIN A PRIZE OF FUTURE PLANTS?

As a crop with a reliable food security, an economic benefit and a high potential of added-value, cassava has drawn high attention in the world (Jansson et al. 2009). A good way to predict the future of cassava may be to measure a tendency of the world production by examining total production of the past twenty years (Figure 3, data from FAO). From 1990 to 2009, the world production has increased dramatically. Interestingly, the major contribution was from Africa and Asia where cassava was relatively newly introduced. The reason may be that many new characteristics and advantages of cassava have been recognized and appreciated in the new farming regions. Secondly, using cassava as a bioenergy crop has recently been admitted, especially in Thailand, China and India (Jansson et al. 2009). This could explain why Asia has made a significant contribution to the increase of cassava world production during the last seven years (Figure 3 A).

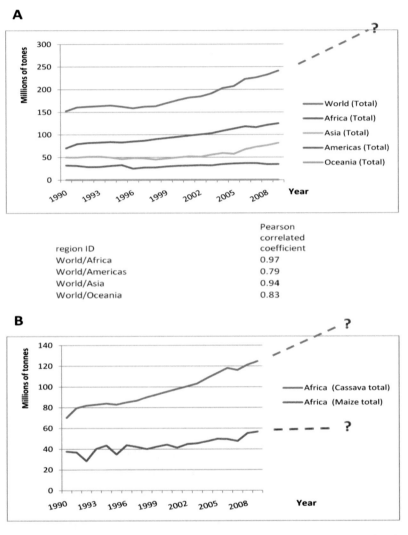

Figure 3. Total world production of cassava in the past twenty years and speculated production in the future. A. Production in the world and on different continents from 1990 to 2009 (solid line) and the possible production in the future (broken line). Correlation between the world and different regions are indicated by a Pearson correlated coefficient in the lower panel. B. Total production of cassava and maize in Africa from 1990 to 2009 (solid lines) and the possible production in the future (broken lines).

If this tendency continues, production of cassava will be over 600 M tons (an amount similar to current rice production) after 20-30 years in the world without even counting some of the possible mile-stone successes such as *"crossing the latitude limitation for cassava farming"* (Figure 3A). In addition, due to the hardiness of cassava to endure high temperatures, cassava may partly replace maize in some parts of Africa after a few decades during the course of global warming (Figure 3 B; http://www.slideshare.net/ ciatdapa/ climate).

ACKNOWLEDGMENTS

Funded by the following organizations and foundations:

- The Swedish Research Council for Environment, Agricultural Sciences and Spatial Planning (Formas) under the Strategic Research Area for the TCBB Program.
- The SLU Lärosätesansökan Program (TC4F) for Team 4 supported by Vinnova.
- The SLU program BarleyFunFood.
- The Swedish International Development Cooperation Agency (Sida/SAREC).
- Joint Formas-Sida funded programs for cassava and rice research, respectively, on sustainable development in developing countries.
- The Carl Trygger Foundation.
- The Chinese International Science and Technology Cooperation Program (2010DFA62040).
- In part by the National Non-profit Institute Research Grant of CATAS-ITBB ZX2008-4−3, China.

We thank Xia Yan for her help in making the graphics of Figure 3.

REFERENCES

Baguma Y., Sun C., Ahlandsberg S., Mutisya J., Palmqvist S., Rubaihayo P.R., Magambo M.J., Egwang T.G., Larsson H. and Jansson C. 2003. Expression patterns of the gene encoding starch branching enzyme II in the storage roots of cassava (*Manihot esculenta* Crantz). *Plant Sci.* 164: 833-839.

Balyejusa Kizito E. B., Bua A., Fregene M., Egwang T., Gullberg U. and A. Westerbergh. 2005. The effect of cassava mosaic disease on the genetic diversity of cassava in Uganda. *Euphytica* 146: 45–54.

Balyejusa Kizito E. B., Chiwona-Karltun L., Egwang T., Fregene M.and Westerbergh A. 2007a. Genetic diversity and variety composition of cassava on small-scale farms in Uganda: An interdisciplinary study using genetic markers and farmer interviews. *Genetica* 130: 301-318.

Baleyjusa Kizito E., Rönnberg-Wästljung A.-C., Egwang T., Gullberg U., Fregene M. and Westerbergh, A. 2007b. Quantitative trait loci controlling cyanogenic glycoside and dry matter content in cassava (Maniho esculenta Crantz). *Hereditas* 144: 129-136.

Bellotti A. and Riis L. 1994. Cassava cyanogenic potential and resitance to pest and diseases. *Acta Hort.* 375: 141-151.

Beyene, D., Baguma, Y., Mukasa, S.B., Sun, C. and Jansson, C. 2010. Characterization and role of isoamylase1 (*MEISA1*) gene in cassava. *African Crop Sci.* J. 18: 1-8.

Burton R.A., Jenner H., Carrangis L., Fahy B., Fincher G.B., Hylton C., Laurie D.A., Parker M., Waite D., van Wegen S., Verhoeven T. and Denyer K. 2002. Starch granule initiation and growth are altered in barley mutants that lack isoamylase activity. *Plant J.* 31: 97-112.

Carter S.E., Fresco L.O., Jones P.G. and Fairbairn J.N. 1992. An atlas of cassava in Africa. Historical, agroecological and demographic aspects of crop distribution. Cali, Colombia, CIAT.

Castelblanco W. and Westerbergh A. 2008. Wild relatives of cassava as genetic sources for improved cassava farmers' varieties and sustainable farming communities. http://www.csduppsala.uu.se/sidaconference08/abstracts/session4.pdf.

Chiwona-Karltun L., Mkumbira J., Saka J., Bovin M., Mahungu N.M. and Rosling H. 1998. The importance of being bitter- a qualitative study on cassava cultivar preference in Malawi. *Ecol. Food Nut.* 37: 219-245.

Chiwona-Karltun L., Brimer L., Saka J.D.K., Mhone A.R., Mkumbira J., Johansson L., Bokanga M., Mahungu N.M. and Rosling H. 2004. Bitter taste in cassava roots correlates with cyanogenic glucoside levels. *J. Sci. Food Agric.* 84: 581-590.

Du L., Bokanga M., Møller B.L. and Halkier B.A. 1995. The biosynthesis of cyanogenic glucosides in roots of cassava. Phytochemistry 39: 323-326.

Elias M., Penet L., Vindry P., McKey D., Panaud O. and Robert T. 2001. Unmanaged sexual reproduction and the dynamics of genetic diversity of a vegetatively propagated crop plant, cassava (*Manihot esculenta* Crantz), in a traditional farming system. *Mol. Ecology* 10: 1895–1907.

El-Sharkawy MA. 2004. Cassava biology and physiology. *Plant Mol. Biol.* 56: 481-501.

Elizabeth E. 2010. Economic analysis of cassava production (farming) in Akwa Ibom State. Agric. *Biol. J. N. Am.* 1: 612-614.

Essers A.J.A., Ebong C., Grift R.M., Nout M.J.R., Otim-Nape W. and Rosling H. 1995. Reducing cassava toxicity by heap-fermentation in Uganda. *Int. J. Food Sci. Nutr.* 46: 125-136.

Fregene M.A., Suarez M., Mkumbira J., Kulembeka H., Ndedya E., Kullaya A., Mitchel S., et al. 2003. Simple sequence repeat marker diversity in cassava landraces: genetic diversity and differentiation in an asexually propagated crop. *Theor. Appl. Genet.* 107: 1083–1093.

Huang Q. and Li J. 2007. Cassava germplasm collection, utilization and breeding. *Guangxi Trop. Agri.* 1: 35-37.

Ihemmere U., Arias-Garzon D., Lawrence S. and Sayre R. 2006. Genetic modification of cassava for enhanced starch production. *Plant Biotechnol. J.* 4: 453-465.

Jansson C., Westerbergh A., Zhang J., Hu X. and Sun C. 2009. Cassava, a potential biofuel crop in China. *Appl. Energy* 86: S95-S99.

Jones W., 1959. Manioc in Africa. Stanford University Press, Stanford.

Jørgensen K., Bak S., Kamp Busk P., Sørensen C., Olsen C.-E., Puonti-Kaerlas J. and Lindberg Møller B. 2005. Cassava plants with a depleted cyanogenic glucoside content in leaves and tubers. Distribution of cyanogenic glucosides, their site of synthesis and transport, and blockage of biosynthesis by RNA interference technology. *Plant Physiol.* 139: 363-374.

Kim H, Ngai NV, Howler R, Ceballos H 2008. Current situation of cassava in Vietnam and its potential as bio-fuel. http://cassavaviet.blogspot.com/2008/09/current-situation-of-cassavain-vietnam.html.

Koch B., Nielsen V.S., Halkier B.A., Olsen C.E. and Møller B.L. 1992. The biosynthesis of cyanogenic glucosides in seedlings of cassava (Manihot esculenta Crantz) *Arch. Biochem. Biophys.* 292: 141-150.

Lancaster P.A. and Brooks J.E. 1983. Cassava leaves as human food. *Econ. Bot.* 37: 331-348.

Li K., Ye J. and Huang J. 2008. Cassava cultivation technology for high yield. Text book for internal training courses of the Chinese Academy of Tropical Agricultural Sciences.

Morante N., Sanchez T., Ceballos H., Calle F., Perez J.C., Egesi C., Cuambe C.E. Escobar A.F. Ortiz D., Chavez A.L. and Fregene M. 2010. Tolerance to post-harvest physiological deterioration in cassava roots. *Crop Sci.* 50: 1333-1338.

Nweke F., Spencer D. and Lynam J. 2002. The cassava transformation: Africa's best-kept secret. Michigan State Univ Press, USA.

Ohlrogge J., Allen D., Berguson B. DellaPenna D., Shachar-Hill Y. And Stymne S. 2009. Driving on biomass. *Science* 324: 1019-1020.

Olsen K.M. and B.A. Schaal. 1999. Evidence on the origin of cassava: phylogeography of *Manihot esculenta*. *Proc. Nat. Acad. Sci. USA* 96: 5586–5591.

Olsen K.M. and B.A. Schaal. 2001. Microsatellite variation in cassava (*Manihot esculenta*, Euphorbiaceae) and its wild relatives: further evidence for a southern Amazonian origin of domestication. *Am. J. Bot.* 88: 131–142.

Padmaja G 1995. Cyanide detoxification in cassava for food and feed uses. *Crit. Rev. Food Sci. Nutr.* 35: 299-339.

Ravindran V. 1993. Cassava leaves as animal feed: potential and limitations. *J. Sci. Food Agri.* 61: 141-150.

Riis L., Bellotti A.C., Bonierbale M. and O'Brien G.M. 2003. Cyanogenic potential in cassava and its influence on a generalist insect herbivore *Cyrtomenus bergi* (Hemiptera: Cydnidae). *J. Econ. Entomol.* 96: 1905-1914.

Rogers D.J. and Appan S.G. 1973. *Manihot*, Manihotoides (Euphorbiaceae). Flora Neotropica. Monograph 13. Hafner, New York, NY.

Schwall G.P., Safford R., Westcott R.J., Jeffcoat R., Tayal A., Shi Y.-C., Gidley M.J. and Jobling S.A. 2000. Production of very-high-amylose potato starch by inhibition of SBE A and B. *Nat. Biotech.* 18: 551-554.

Siritunga D., Arias-Garzon D., White W. and Sayre R.T. 2004. Over-expression of hydroxynitrile lyase in cassava roots accelerates cyanogenesis and detoxification. *Plant Biotech. J* 2: 37-43.

Siritunga D. and Sayre R.T. 2003. Generation of Cyanogen-Free Transgenic Cassava. *Planta* 217: 367-373.

Taylor N., Chavarriaga P., Raemakers K., Siritunga D. and Zhang P. 2004. Development and application of transgenic technologies in cassava. *Plant Mol. Biol.* 56: 671-688.

Turesson, H., Marttila, S., Gustavsson, K.-E., Hofvander, P., Olsson, M. E., Bulow, L., Stymne, S., Carlsson, A.S. 2010. Characterization of oil and starch accumulation in tubers of Cyperus esculentus var. sativus (Cyperaceae): A novel model system to study oil reserves in non-seed tissues. *Am. J. Bot.* 97: 1884-1893

Tylleskär T., Banea M., Bikangi N., Cooke R., Poulter N. and Rosling H. 1992. Cassava cyanogens and konzo, an upper motor neuron disease found in Africa. *Lancet* 339: 208-211.

Wang D., Portis A.R.Jr., Moose S.P.and Long S.P. 2008. Cool C4 photosynthesis: pyruvate Pi dikinase expression and activity corresponds to the exceptional cold tolerance of carbon assimilation in Miscanthus x giganteus. *Plant Physiol.* 148: 557-567.

White W., McMahon J. and Sayre R. 1994. Regulation of cyanogenesis in cassava. *Acta Hort.* 375: 69-77.

Zhang P., Jaynes J.M., Potrykus I., Gruissem W. And Puonti-Kaerlas J. 2003. Transfer and expression of an artificial storage protein (ASP1) gene in cassava (Manihot esculenta Crantz). 12: 243-250.

In: Cassava: Farming, Uses, and Economic Impact   ISBN: 978-1-61209-655-1
Editor: Colleen M. Pace   © 2012 Nova Science Publishers, Inc.

Chapter 8

NEUROLOGICAL DISORDERS ASSOCIATED WITH CYANOGENIC GLYCOSIDES IN CASSAVA: A REVIEW OF PUTATIVE ETIOLOGIC MECHANISMS

*Bola Adamolekun**

Department of Neurology, University of Tennessee Health
Science Center, Memphis, TN, U.S.A.

ABSTRACT

Tropical ataxic neuropathy (TAN) and konzo are two neurological disorders associated with the chronic consumption of cassava (*Manihot esculenta)* in several African countries. TAN is characterized by sensory polyneuropathy, sensory ataxia, bilateral optic atrophy and bilateral sensori-neural deafness. It occurs in poor, undernourished elderly individuals subsisting on a monotonous cassava diet with minimal protein supplementation. Konzo is a syndrome of symmetrical, non-remitting, non-progressive spastic paraparesis, with a predilection for children and women of child-bearing age. It is invariably associated with monotonous consumption of inadequately processed bitter cassava roots with very minimal protein supplementation.

* Address for correspondence: B. Adamolekun, MD, FWACP, Department of Neurology, University of Tennessee Health Science Center, 855 Monroe Avenue, Memphis TN 38163, USA. Phone: 901-4484916. Fax: 901-4487440. badamole@uthsc.edu.

Chronic cyanide intoxication from consumption of cyanogenic glycosides in cassava was long thought to be the major etiological factor for TAN, but there has been no evidence of a causal association. Similarly, high cyanide consumption with low dietary sulfur intake due to almost exclusive consumption of insufficiently processed bitter cassava roots was proposed as the cause of konzo, but there has also been no evidence of a causal association. The roles of the cyanogenic glycoside linamarin, acetone cyanohydrin and cyanate in the etiology of konzo had been evaluated, but there was no evidence of a causal association. The etiologies of both TAN and konzo therefore remained unknown, despite studies in several countries aimed at unraveling the etiologic mechanisms of these debilitating diseases.

In this chapter an etiological mechanism of thiamine deficiency for both TAN and konzo is discussed. It is postulated that in TAN and konzo patients, thiamine deficiency results from the inactivation of thiamine that occurs when, in the absence of dietary sulfur-containing amino acids in these patients with poor protein intake; the sulfur in thiamine is utilized for the detoxification of cyanide in the cyanogenic glycosides consumed in cassava. Thiamine is known to be rendered inactive when the sulfur in its thiazole moiety is combined with hydrogen cyanide. Evidence from the literature implicating chronic thiamine deficiency in the etiology of TAN are discussed. These include evidence of abnormal pyruvate metabolism reversed by thiamine in patients with TAN, evidence from erythrocyte transketolase activity indicating significant thiamine deficiency in patients with TAN compared to controls, and a placebo-controlled trial of therapeutic doses of thiamine which showed a clinically dramatic and statistically significant improvement in ataxia.

Thiamine status has never been evaluated in patients with konzo, and a therapeutic trial of thiamine has not been conducted. Evidences in support of thiamine deficiency as the etiological mechanism of konzo include a demonstrated evidence of widespread thiamine deficiency in the susceptible population, animal studies demonstrating the similarity of the clinical presentations of konzo to those of thiamine deficiency, and the predilection of konzo for adolescent children and women of child-bearing age; population groups particularly vulnerable to symptomatic thiamine deficiency in the presence of inadequate thiamine intake. Studies are currently underway to confirm the role of thiamine deficiency in the etiologies of these debilitating neurological disorders.

INTRODUCTION

The processed roots and leaves of Cassava (*Manihot esculenta)* are eaten by an estimated half a billion people in the tropics and subtropics. Almost half of the current world production of cassava takes place in Africa [1]. Since it can withstand poor rainfall and it yields well in poor soil, cassava is often the only food available to be consumed during drought conditions in several communities in Africa.

Cassava is a cyanophoric plant producing linamarin, a cyanogenic glycoside present in the leaves and roots of the plant. The enzyme linamarase catalyzes the hydrolysis of linamarin to form acetone cyanohydrin, which is hydrolyzed to hydrogen cyanide and acetone. The cyanide is then detoxified by conversion to thiocyanate, a reaction that involves sulfur as a rate-limiting co-factor for the enzyme rhodanese [2]. The concentration of sulfur is dependent on the availability of the sulfur amino acid methionine from dietary protein. Thiocyanate is excreted in the urine. Cyanide may also be metabolized to cyanate, which can be converted by the enzyme cyanase to ammonia and bicarbonate [2].

Although cassava roots are rich in calories, they are markedly deficient in proteins, particularly the essential sulfur amino acid methionine [3]. Cassava is also known to be deficient in thiamine, niacin and riboflavin [4].

Tropical ataxic neuropathy (TAN) and konzo are two neurological disorders occurring in impoverished patients in western, eastern and southern Africa. These disorders have been associated with the chronic, monotonous consumption of cassava associated with very minimal protein supplementation; and will be discussed in detail in this chapter.

EPIDEMIOLOGY OF NEUROLOGICAL DISORDERS ASSOCIATED WITH CASSAVA CONSUMPTION

Tropical Ataxic Neuropathy (TAN)

Tropical ataxic neuropathy (TAN) is a syndrome of sensory polyneuropathy, sensory ataxia, bilateral optic atrophy and bilateral sensori-neural deafness. The syndrome has been described in Nigeria [5, 6], Tanzania [7], Sierra Leone [8] and India [9]. In an epidemiological study of 206 patients with TAN [10], 98% had a sensory polyneuropathy and 84 % had a sensory gait ataxia with a positive Rhomberg's sign. Optic neuropathy was present in 48% and sensori-neural deafness in 19%. The prevalence of TAN increases with age, with a peak in the 5th and 6th decades of life. The illness is initially progressive and later becomes static [11].

Konzo

Konzo is an upper motor neuron disease, characterized by abrupt onset of symmetrical, non-progressive, non-remitting spastic paraparesis [12]. Epidemics have been reported from Tanzania [13], from the Central African Republic [14], from Mozambique [15] and from Democratic Republic of Congo [16]. Thousands of cases of konzo have been reported from these countries, and the disease still continues to be prevalent [17].

The diet in patients who succumb to konzo in all countries from where it has been described consists of monotonous consumption of improperly processed bitter cassava roots, with very minimal protein supplementation; often associated with famine or war.

Unlike TAN which is a disease of the elderly, konzo preferentially affects children between 4 and 12 years and women of reproductive age [18]. It is the most common cause of gait disability in these age groups in the affected areas. The least affected areas in endemic countries are the coastal zones where the availability of fish leads to improved protein supplementation of the cassava diet; and the district capitals with access to commercially available variety of foods [18].

The clinical hallmark of konzo is the abrupt onset of symmetrical, non-progressive, non-remitting spastic paraparesis presenting as gait difficulties of varying severity [12].

Another clinical presentation is optic neuropathy. A neuro-ophthalmology evaluation showed that up to 52% of konzo patients had symptoms qualifying for the diagnosis of optic neuropathy [19]. In konzo patients there may be atrophy of the papillo-macular nerve fiber layer with temporal disc pallor, findings characteristic of optic neuropathy [12]. In one study, visual evoked potentials were abnormal in 48% of konzo patients, indicating the presence of axonal loss in the pre-chiasmal visual pathways, diagnostic of optic neuropathy [20].

Peripheral sensory neuropathy is common in konzo patients. In one study, all konzo patients had symptoms of painful dysesthesias in the lower limbs, prior to onset of spastic paraparesis [21].

Somatosensory evoked potential studies indicated prolongation or absence of cortical responses, confirming involvement of the somatosensory pathways in 60% of konzo patients studied in one series, and 100% of patients studied in another series [22].

PUTATIVE ETIOLOGIC MECHANISMS FOR TAN

The Cyanide Hypothesis

It has long been suggested that chronic cyanide intoxication from cassava meals is a major etiological factor in TAN [5, 9, 23], based on the history of almost total dependence on a monotonous diet of cassava derivatives in patients with the syndrome. This view appeared to have been supported by the results of field surveys which showed a high TAN prevalence in cassava-eating villages, whereas the disease did not exist in a village where cassava was not the predominant staple food [5]. High serum thiocyanate levels have been demonstrated in patients with TAN [24, 25]. Plasma and urinary thiocyanate levels have been shown to fall when TAN patients were fed on cassava-free diets but rose again when patients reverted to a cassava-containing diet [5, 25] These studies of thiocyanate levels, while supporting the fact that eating cassava results in exposure to cyanide which is detoxified to thiocyanate; did not indicate that cyanide toxicity was the cause of TAN. High levels of thiocyanate excretion in patients with TAN are more indicative of effective detoxification of cyanide than of increased risk of toxicity from cyanide.

The results of a study in Nigeria showing a very low prevalence of TAN in a community where exposure to cyanide from cassava consumption was high [26] indicated that exposure to cyanide from cassava was unlikely to be the cause of the disease. A case-control study also did not show an association between exposure to cyanide from cassava foods and TAN [27], and the level of intake of cassava foods is known to be similar in cases of TAN and their unaffected family members [5] and controls [27].

TAN has been reported from communities where cassava is not the staple food. In a report of endemic TAN in Tanzania, the staple diet of the affected individuals was not cassava but rice [7]. The hypothesis that chronic cyanide intoxication was the etiological factor in TAN was tested in two double-blind, placebo-controlled studies of hydroxocobalamin, a potent cyanide antagonist. Both studies demonstrated no beneficial effect [28, 29].

Sulfur Amino Acid Deficiency

Sulfur-containing amino acids are required for the detoxification of cyanide to thiocyanate. An investigation of 9 patients with TAN showed an absence or marked diminution of plasma levels of the sulfur - containing amino acids cysteine and methionine [30], raising the possibility that low levels of sulfur –containing amino acids may be an etiological factor for TAN. However, a double-blind, placebo-controlled study of cysteine in patients with TAN did not demonstrate any beneficial effect [29]

PUTATIVE ETIOLOGICAL MECHANISMS FOR KONZO

The Cyanide Hypothesis

The cyanide hypothesis suggested that konzo is caused by cyanide intoxication from insufficiently processed bitter cassava in combination with a sulfur amino acid–deficient diet [31, 32], with the high cyanide intake indicated by high serum and urinary thiocyanate levels. However, spastic paraparesis, the clinical hall mark of konzo is not a known clinical manifestation of cyanide toxicity; and has not been associated with cyanide exposure from any other source. Further, patients with konzo and their family members who did not succumb to the disease have been shown to be exposed to similar high levels of cassava consumption, with both groups having high thiocyanate levels with no correlation between disease severity and thiocyanate level [18].

Lastly, high serum and urinary thiocyanate levels, while supporting the fact that the consumption of inadequately processed bitter cassava results in exposure to cyanide which is detoxified to thiocyanate, did not indicate that cyanide toxicity was the cause of Konzo. Indeed, high levels of thiocyanate excretion are more indicative of effective detoxification of cyanide than of increased risk of toxicity from cyanide.

Acetone Cyanohydrin Hypothesis

Since Konzo does not present with the known clinical effects of cyanide, it was hypothesized that perhaps acetone cyanohydrin (the aglycone of linamarin) may be the cause of Konzo [33]. However, a study designed to test this hypothesis showed that rats exposed to acetone cyanohydrin did not show any persistent motor deficits that are comparable to konzo [33].

Cyanate Hypothesis

Cyanide from cyanogenic glycosides in cassava may be metabolized to cyanate. There is a significant increase in the plasma cyanate concentrations of sulfur amino acid –deficient rats treated with potassium cyanide [2]. Prolonged cyanate treatment is known to induce neuropathologic abnormalities [34], but there is no evidence that such neuropathologic abnormalities are compatible with TAN or konzo.

Linamarin Hypothesis

It has been suggested that a specific neurotoxic effect of the cyanogenic glycoside linamarin, rather than the associated general cyanide exposure may be the cause of konzo, with a study suggested that linamarin could be transported to the cytoplasm of neural cells where it could cause degeneration [35]. There is however no evidence to suggest that putative neurodegeneration by linamarin may lead to the motor deficits seen in konzo.

Nitrile Hypothesis

It has recently been suggested that TAN and konzo may be caused by different but similar nitriles via neurotoxic actions independent of systemic cyanide release [36]. The nitriles are thought to be either present in cassava or generated during food processing or in the human body. While several small nitriles can cause a variety of neurotoxic effects, no evidence has been presented to indicate that nitriles can cause the clinical features seen in TAN and konzo.

ETIOLOGICAL MECHANISM OF THIAMINE DEFICIENCY FOR TAN AND KONZO

An etiological mechanism of thiamine deficiency has been suggested for TAN [37] and konzo [38]. It is posited that TAN is a chronic thiamine deficiency state occurring in impoverished African communities that consume a monotonous diet of cassava with minimal protein supplementation. TAN results from the chronic thiamine deficiency that is the consequence of poor dietary thiamine intake and the inactivation of thiamine by cyanogenic glycosides in cassava [37].

It is also postulated that konzo is a thiamine deficiency state resulting from the inactivation of thiamine that occurs when, in the absence of dietary sulfur-containing amino acids; the sulfur in thiamine is utilized for the detoxification of cyanide consumed in improperly processed bitter cassava [38].

Putative Mechanism of Thiamine Deficiency

Following consumption of cyanogenic glycosides in cassava, cyanide is converted to thiocyanate by the enzyme rhodanese, a reaction that involves sulfur as a rate-limiting co-factor [2]. The concentration of sulfur is normally dependent on the availability of sulfur amino acids from dietary proteins. However, rhodanese has wide substrate specificity, with respect to both sulfur donor and acceptor. Therefore, sulfur compounds other than sulfur amino acids may function as sulfur donors [39]. Available sulfur is preferentially utilized for cyanide intoxication, even in protein malnutrition. In konzo patients, a low urinary sulfur concentration in the presence of high serum thiocyanate levels indicated dietary sulfur shortage and the mobilization of endogenous sulfur from sulfur compounds other than sulfur amino acids [32].

Thiamine is a sulfur compound whose sulfur can be mobilized during shortage of dietary sulfur amino acids. The thiamine molecule is composed of a pyrimidine moiety and a sulfur-containing thiazole moiety. Thiamine is known to be rendered inactive when the sulfur in the thiazole moiety is combined with hydrogen cyanide [40].

EVIDENCE IN SUPPORT OF THIAMINE DEFICIENCY

There is widespread thiamine deficiency in TAN and konzo- susceptible populations.

The cassava-induced inactivation of thiamine is more likely to lead to clinical thiamine deficiency syndromes in populations with marginal thiamine intake and low baseline thiamine levels. This appears to be the case in an endemic area for TAN, where a nutritional survey in western Nigeria indicated a deficiency of thiamine intake in the rural and urban population of all ages [41]. It is also the case in Democratic Republic of Congo (DRC, an endemic country for konzo), where there is evidence of low thiamine intake below 60% of dietary reference intake [42]. There were also low baseline thiamine levels assessed by determination of the activation coefficient of erythrocyte transketolase, which showed that 51 % of apparently healthy residents in DRC had values compatible with a low or deficient thiamine status [42].

Clinical features of TAN and konzo are compatible with thiamine deficiency.

Optic neuropathy, a clinical manifestation of both TAN and konzo, is known to result from a diet-induced deficiency of thiamine [43]. Studies in animals have also demonstrated optic nerve degeneration with thiamine deficiency [44].

Sensory polyneuropathy, a clinical feature of both TAN and konzo; and the resulting gait ataxia are well known presentations of dry beriberi seen in thiamine deficiency [45]. Indeed, the Save the Children Fund (UK) case screening definition for thiamine deficiency [45,46] included sensory polyneuropathy, ataxia and impairment of vision or hearing; all clinical features seen in patients with TAN.

Spastic paraparesis, the clinical hallmark of konzo, may occur with thiamine deficiency. Severe thiamine deficiency has been shown to result in spastic paraparesis in foxes, dogs and cattle. Chastek paralysis, characterized by spastic papaparesis in foxes, was found to be due to thiamine deficiency induced by thiaminases in raw fish fed to the foxes [47, 48]. The syndrome of polioencephalomalacia, characterized by anorexia and spastic paraparesis, has been reported in dogs that developed thiamine deficiency from eating over-cooked meat [49]. The clinical symptoms resolved with thiamine hydrochloride. The syndrome of polio-encephalomalacia has also been reported in thiamine –deficient cattle [50]. Umoh et al (1985) studied rats fed a no-thiamin diet for 40 days, compared with rats fed a low thiamin and a normal thiamin diet [51]. During the feeding period, the rats receiving no thiamine developed stiffness of the hind legs, akin to the lower limb spasticity in patients with konzo.

Therapeutic Trial of Thiamine in TAN

Thiamine is required for pyruvate metabolism, and a deficiency of thiamine results in increased blood pyruvate levels which can be reversed by thiamine supplementation. Monekosso showed a disturbance of pyruvate metabolism in Nigerian patients with TAN, which was reversed by vitamin B complex, a multi-vitamin preparation containing 100mg of thiamine [7, 52]. These findings prompted a controlled therapeutic trial of vitamin B complex in hospitalized patients with TAN [52]. The choice of vitamin B complex rather than thiamine alone was impelled by the author's concern that 100mg of thiamine, if given alone could precipitate a "painful feet syndrome" in the patients. Apart from thiamine, the other contents of vitamin B complex are pyridoxine, nicotinamide (both of which are not deficient in patients with TAN [53]) and riboflavin (which did not show any therapeutic benefit in a controlled trial in TAN [28]).

At the end of 7 days of therapy, the mean ataxic scores in the treatment group showed a dramatic and statistically significant improvement, compared to the control group [52]. At 2 weeks, there was no longer a statistically significant difference in ataxia scores, as TAN patients in the control group had slowly improved, presumably because of the relatively rich thiamine content of hospital food. The improvement in ataxia scores at 7 days is similar to the rapid therapeutic response to thiamine therapy reported by this author in another ataxic syndrome in western Nigeria [54] that is known to be secondary to thiamine deficiency [54,55].

There have been no studies of thiamine status in patients with konzo, and no clinical trials of thiamine for the prevention or therapy of the syndrome.

DISCUSSION

In this chapter, evidence has been presented indicating that tropical ataxic neuropathy (TAN) is a thiamine-deficiency state. These include evidence of abnormal pyruvate metabolism reversed by thiamine in patients with TAN [7, 53], evidence from erythrocyte transketolase activity indicating significant thiamine deficiency in patients with TAN compared to controls [53], and a placebo-controlled trial of therapeutic doses of thiamine which showed a clinically dramatic and statistically significant improvement in ataxia [53]. The clinical presentation of TAN has also been shown to be compatible with thiamine deficiency [37]. The peak age of presentation, clinical course and the clinical presentation of TAN appear to be compatible with those of an insidious, smoldering type of thiamine deficiency [37].

Konzo has also been postulated to be a thiamine deficiency state [38]. The putative mechanism for thiamine deficiency in konzo is credible because it integrates the core epidemiological events in konzo, which are consistently found in geographically disparate parts of Africa over different periods of time. These include the monotonous cassava diet, the insufficient cassava processing, the high intake and detoxification of cyanogenic glycosides indicated by high urinary thiocyanate levels, and the low intake of sulfur-containing amino acids, demonstrated by low urinary sulfate excretion [2, 32]. It explains why konzo has not been described or reported to occur from cyanide exposure without simultaneous malnutrition. Importantly, the clinical features of konzo are shown to be compatible with those of thiamine deficiency [38].

The mean serum thiocyanate levels in patients with konzo are considerably higher than those found in patients with TAN [31] and the blood cyanide levels found in konzo patients at onset of disease is 20 times higher than levels in TAN patients [31]. Konzo patients thus appear to consume much higher levels of cyanogenic glycosides compared to konzo patients, and this may account for the more, fulminant presentation of konzo.

The severity of thiamine deficiency is known to be important in determining the symptomatology and clinical presentation of thiamine deficiency. Pigeons allowed to feed voluntarily on thiamine-deficient diets will unpredictably become acutely (opisthotonus) or chronically (leg weakness, ataxia) thiamine-deficient, depending on how much they eat and the amount of thiamine they consume [56]. The presence of optic neuropathy in 52% of konzo patients [19] and of peripheral sensory neuropathy in 60-100% of konzo patients [22], percentages similar to those in TAN patients [10]; suggest that TAN may actually be present in konzo-affected populations.

Programs to prevent TAN and konzo have focused on producing and distributing less toxic varieties of cassava and disseminating new processing methods, such as grating and the flour wetting method. Despite these, epidemics still continue to occur [57]. A long –term thiamine-supplementation program for susceptible populations in the endemic areas may be more effective in the control and eventual eradication of these debilitating diseases.

REFERENCES

[1] Nhassico D, Muquingue H, Cliff J, Cumbana A, Bradbury JH. Rising African cassava production, diseases due to cyanide intake and control measures. *J. Sci. Food Agric.* 2008; 88: 2043-9.

[2] Tor-Agbidye J, Palmer VS, Lasarev MR, Craig MA, Blythe LL, Sabri MI, Spencer PS. Bioactivation of cyanide to cyanate in sulfur amino acid deficiency: relevance to neurologic disease in humans subsisting on cassava. *Toxicological Sciences* 1999. 50: 228-235.

[3] Adegbola AA: Methionine as an addition to cassava-based diets. : In: B Nestel and M Graham (eds). Cassava as animal feed: Proceedings of a workshopheld at the University of Guelph. IDRC Ottawa, Canada 1977. 9-17.

[4] Latham MC. Human Nutrition in tropical Africa FAO Rome 1969.

[5] Osuntokun BO. An ataxic neuropathy in Nigeria: A clinical, biochemical and electrophysiological study. *Brain* 1968; 91: 215- 48.

[6] Monekosso GL, Annan WGT. Clinical epidemiological observations on an ataxic syndrome in Western Nigeria. *Trop Geogr. Med.* 1964; 4: 316-23.

[7] Haddock, DRW, Ebrahim GJ, Kapur B. Ataxic Neurological syndrome found in Tangayika. *Brit. Med. J.* 1962. (2): 1442-1443.

[8] Rowland H, Neuropathy in Sierra Leone. *J. Trop. Med. Hyg.*, 1963, 66: 181-187.

[9] Madhusudanam M, Menon MK, Ummer K, Radhakrishnanan K. Clinical and etiological profile of tropical ataxic neuropathy in Kerala, South India. *Eur. Neurology*, 2008. 60 (1): 21-26.

[10] Oluwole OSA, Onabolu AO, Link H et al. Persistence of tropical ataxic neuropathy in a Nigerian community. *J. Neurol. Neurosurg. Psychiatry* 2000; 69: 96-101.

[11] Osuntokun BO. Chronic cyanide intoxication of dietary origin and a degenerative neuropathy in Nigerians. Acta Horticulturae 1994. 375; 271-83.

[12] Tylleskar T, Howlett WP, Rwiza HT, Aquilonius S-M, et al. Konzo: a distinct disease entity with a selective upper motor neuron damage. *J. Neurol. Neurosurg. Psychiatry*; 1993: 56: 638-643.

[13] Howlett, WP, Brubaker GR, Mlingi N, Rosling H. Konzo, an epidemic upper motor neuron disease studied in Tanzania. *Brain.* 1990 113 (1). 223-35.

[14] Tylleskar T, Legue FD, Peterson S, Kpizingui E, Stecker P. Konzo in the Central African Republic. *Neurology* 1994; 44:959.

[15] Cliff J, Nicala D, Saute F, Givragy R, Azambuja G, Taela A et al. Konzo associated with war in Mozambique. *Trop. Med. Int. Health* 2: 1068-1074. 1997.

[16] Tylleskar T, Banea M, Bikangi N, Fresco L, Persson LA, Rosling H. Epidemiological evidence from Zaire for a dietary etiology of konzo, an upper motor neurone disease. *Bull. World Health Organ.* 1991: 69 (5): 581-589.

[17] Earnesto M, Cardoso AP, Nicala D, Mirione E, Massaza F, Cliff J et al. Persistent konzo and cyanide toxicity from cassava in northern Mozambique. *Acta Trop.* 2002; 82: 357-362.

[18] Ministry of Health, Mozambique. Mantakassa: an epidemic of spastic paraparesis associated with chronic cyanide intoxication in a cassava staple area of Mozambique. 1. Epidemiology and clinical laboratory findings in patients. *Bull. World Health Organ.* 1984a. 62 (3): 477-484.

[19] Mwanza JC, Tshala-katumbay D, Kayembe DL, Eeg-Olofsson KE, Tylleskar T. Neuro-ophthalmologic findings in konzo, an upper motor neurone disorder in Africa. *Eur. J. Ophthalmol.* 2003: 13 (4): 383-9.

[20] Mwanza JC, Lysebo DE, Kayembe DL, Tshala-katumbay D, Nyamabo LK, Tylleskar T, Plant GT. Visual evoked potentials in konzo, a spastic paraparesis of acute onset in Africa. *Ophthalmologica* 2003; 217 (6): 81-6.

[21] Carton H, Kazadi K, Kabeya, Odio et al. Epidemic spastic paraparesis in Bandundu (Zaire). *J. Neurol. Neurosurg. Psychiatry* 1986; 49: 620-7.

[22] Tshala-Katumbay D, Eeg-Olofsson K, Kazadi-Kayembe F, Fallmar P, Tylleskar T, Kayembe_Kalula T. Abnormalities of somatosensory evoked potentials in Konzo- An upper motor neuron disorder. *Clin. Neurophysiol.* 2002. 113 (1): 10-15.

[23] Osuntokun BO, Monekosso GL, Wilson J. Relationship of a degenerative tropical Neuropathy to diet: Report of a field survey. *Brit. Med. J.* 1969 1: 547-550.

[24] Monekosso, GL and Wilson J. Plasma thiocyanate and vitamin B1 in Nigerian patients with degenerative neurological disease. *Lancet* 1966. 1: 1062-1064.

[25] Osuntokun BO, Monekosso GL, Wilson J. Relationship of a degenerative tropical Neuropathy to diet: Report of a field survey. *Brit. Med. J.* 1969 1: 547-550.

[26] Oluwole OSA, Onabolu AO, Cotgreave IA, et al. Low prevalence of ataxic polyneuropathy in a community with high exposure to cyanide from cassava foods. *J. Neurol.* 2002; 49: 1034-40.

[27] Oluwole OSA, Onabolu AO, Cotgreave IA, et al. Incidence of endemic ataxic polyneuropathy and its relation to exposure to cyanide in a Nigerian community. *J. Neurol. Neurosurg. Psychiatry* 2003; 74: 1417-1422.

[28] Osuntokun BO, Langman MJS, Wilson J, Aladetoyinbo A. Controlled trial of hydroxocobalamin and riboflavine in Nigerian ataxic neuropathy. *J. Neurol. Neurosurg. Psychiat.* 1970; 33: 663-666.

[29] Osuntokun BO, Langman MJS, Wilson J, et al. Controlled trial of combinations of hydroxocobalamin-cystine and riboflavine-cystine in Nigerian ataxic neuropathy. *J. Neurol. Neurosurg. Psychiat.* 1974. 37, 102-104.

[30] Osuntokun, BO, Durowoju JE, McFarlane H, Wilson J et al. Plasma amino acids in the Nigerian Nutritional ataxic neuropathy. *Brit. Med. J.* 1968 3: 647-649.

[31] Tylleskar, T. The association between cassava and the paralytic disease konzo. *Acta Horticulturae* 1994. 375: 331-339.

[32] Cliff J, Lundquist P, Martenson J, Rosling H, Sorbo B. Association of high cyanide and low sulfur intake in cassava –induced spastic paraparesis. *Lancet* 1985; 1211-1213.

[33] Soler-Martin C, Riera J, Seoane A, Cutillas B, Ambrosio S, Boadas-Vaello P, Llorens J. The targets of acetone cyanohydrin neurotoxicity in the rat are not the ones expected in an animal model of konzo. *Neurotoxicol Teratol.* 2009.

[34] Tellez I, Johnson D, Nagel RL, Cerami A. Neurotoxicology of sodium cyanate: New pathological and ultrastructural observations in Macaca nemestrina. *Acta neuropathol.* 1979. 47: (1): 75-79.

[35] Sreeja VG, Nagahara N, Li Q, Minami M. New aspects in pathogenesis of konzo: neural cell damage directly caused by linamarin contained in cassava (Manihot esculenta Crantz). *Brit. J. Nutr.* 2003. 90; 467-472.

[36] Llorens J, Soler-Martín C, Saldaña-Ruíz S, Cutillas B, Ambrosio S, Boadas-Vaello P. A new unifying hypothesis for lathyrism, konzo and tropical ataxic neuropathy: Nitriles are the causative agents.*Food Chem. Toxicol.* 2010 Jun 8. [Epub ahead of print].

[37] Adamolekun B. Thiamine deficiency and the etiology of tropical ataxic neuropathy. *International Health* 2010. 2 (1): 17-21.

[38] Adamolekun B. Etiology of konzo, epidemic spastic paraparesis associated with cyanogenic glycosides in cassava: Role of thiamine deficiency? *J. Neurol. Sci.* 2010. 296 (1-2): 30-33.

[39] Westley J: Thiosulfate: cyanide sulfur transferase (rhodanese). *Methods Enzymol.* 1981. 77; 85-91.

[40] McCandless, DW. African Seasonal Ataxia. In: McCandless DW: Thiamine deficiency and associated clinical disorders 2009. Humana press, NY :103-112.

[41] Fashakin JB, Oyekanmi M. Evaluation of some vitamin B complex nutriture in Ile-Ife and environs (Nigeria). *J. Nutr. Res.* 1986; 56: 79-84.

[42] Barclay DV, Mauron J, Blondel A, Cavadini C, Vereilghen AM, Van Geert, Dirren H. Micronutrient intake and status in rural Democratic Republic of Congo. *Nutr. Res.* 2003; 3: 659-671.

[43] Hoyt CS, Billson FA. Vision loss on low Carb diets. Low carbohydrate diet optic neuropathy. *The Medical Journal of Australia* 1997 1: 65.

[44] Rodger FC. Experimental thiamine deficiency as a cause of degeneration in the visual pathway of the rat, *Brit. J. Ophthal.*, 1953, 37: 11-17.

[45] World Health Organization, 1999. Thiamine deficiency and its prevention and control in major emergencies. WHO/NHD/99.13.

[46] SCF (UK). Report on the outbreak of suspected thiamine deficiency. Jhapa, Nepal, Bhutanese Refugee Health Project. Save the Children Fund (UK), 1994.

[47] Green RG, Carlson WE and Evans CA. A deficiency disease of foxes produced by feeding fish. *J. Nutr.* 1941; 21: 243-356.

[48] Yudkin WH. Thiaminase, the Chastek-paralysis factor. *Physiol. Rev.* 1949: 388-402.

[49] Read DH, Jolly RD, Alley MR. Polioencephalomalacia of dogs with thiamine deficiency *Vet. Pathol.* 1977 14 (2): 103-12.

[50] Loew FM, Bettany JM, Halifax CE. Apparent thiamine status of cattle and its relationship to polioencephalomalacia. *Can. J. Comp. Med.* 1975 39: 291-5.

[51] Umoh IB, Ogunkoya FO, Maduagwu EN, Oke OL. Effect of thiamine status on the metabolism of linamarin. *Ann. Nutr. Metab.* 1985. 29: 319-324.

[52] Monekosso GL, Annn WG, Ashby PH. Therapeutic effect of Vitamin B complex on an ataxic syndrome in Western Nigeria. *Trans Roy. Soc. Trop. Med. Hyg.* 1964; 58: 432-6.

[53] Osuntokun BO, Aladetoyinbo A, Bademosi O. Vitamin B nutrition in the Nigerian tropical ataxic neuropathy. *J. Neurol. Neurosurg. Psychiat.* 1985; 48: 154-156.

[54] Adamolekun B, Adamolekun WE, Sonibare AD, Sofowora G. A double-blind, placebo-controlled study of the efficacy of thiamine hydrochloride in a seasonal ataxia in Nigerians. *Neurology* 1994. 44 : 549-551.

[55] Nishimune T, Watanabe Y, Okazaki H, Akai H. Thiamin is decomposed due to Anaphe spp entomophagy in seasonal ataxia patients in Nigeria. *J. Nutrition.* 130; 6: 1625-1628.

[56] Swank Rl, Bessey OA. Avian thiamine deficiency. Characteristic symptoms and their pathogenesis. *J. Nutrition* 1941. 22; 77-89.

[57] Cliff J, Muquingue H, Nhassico D, Nzwalo H, Bradbury JH. Konzo and continuing cyanide intoxication from cassava in Mozambique. *Food Chem. Toxicol.* 2010 Jul 21. [Epub ahead of print].

In: Cassava: Farming, Uses, and Economic Impact   ISBN: 978-1-61209-655-1
Editor: Colleen M. Pace   © 2011 Nova Science Publishers, Inc.

Chapter 9

THE CASSAVA α1, 4 α1, 6 GLUCOPOLYSACCHARIDE

Elda María Salmoral

Group of Biochemical Engineering (G.I.B.),
School of Engineering, University of Buenos Aires,
Paseo Colón 850 (1063), Buenos Aires, Argentina

ABSTRACT

Commonly known as starch, the α1, 4 α1, 6 glucopolisaccharide depends on its amylose: amylopectin ratio to understand their physicochemical properties. Amylose is considered a long linear polymer with D-glucosyl units linked through by α D-1, 4 glucose linkages and amylopectin has a branched structure with poly-glucose residues linked by α D-1, 4 and α D-1, 6 glycose linkages.

As an important prerequisite to evaluate the cassava starch granule, a carefully isolation of the starch in absence of a non degradative process is considered in this chapter in laboratory scale.

The performance of the optimum starch molecular fractionation method gives an opportunity to obtain a best knowledge of its components. To determine the efficiency of fractionation starch two methods are presented, one method based on the differential solubility of starch components in a water- butanol saturated solution and another method based on complexing the starch molecule with Concanavalin A by its high affinity with carbohydrates.

The composition and physical parameters give rise to diverse processing properties and therefore many uses of starch in non-food industries. In this chapter the cassava starch modification is described: 1) by gamma radiation to improve the water capacity property of a plastic biodegradable material, and 2) by esterification with fatty acid chloride to obtain a non polar agent with emulsifying capacity.

INTRODUCTION

The starch is a glucopolisaccharide discovered more than a century ago and it has not stopped being studied until now and while the analysis of their structure in different vegetal species is known, new knowledges arise about its content and its architecture. It can be attributed to overexpression of the enzymes involved in the synthesis of this polymer according to Preiss, J., 2009.

The synthesized starch is storaged in granules placed in the amyloplasts during the maduration and growth of seeds, tuber and roots. Those granules vary in their size and shape depending from one species to another.

Amylose and amylopectin are the two major glucopolysaccharide components of starch granule.

Over the past three decades some models of amylopectin and amylose structure have been proposed suggesting physical aspects and characteristic properties of the starch (Matheson, N., 1996; Robin, J., 1974; Manners, D., 1989; Takeda, Y., 1988; Morrison, W., 1983).

Amylose is a long linear polymer with D-glucosyl units linked through by α D-1, 4 glycosidic linkages, although there is evidence that amylose is not completely linear (Curá, A., 1995). A molecular weight of 10^5 and 10^6 g mol-[1] is attributed to it and has chains of about 840 to 22000 α D-glucopyranosyl units (Roger, P., 1996).

Amylose content is determined by iodine-binding based on amperometric, spectrophotometric, or potentiometric methods (Larson, B., 1953; Karve, M., 1992; Morrison, W., 1983, Banks, W., 1974) or by considering the different number of non-reducing end-groups between amylose and amylopectin (Tolmasky, D., 1987).

Amylopectin, the other polymer has a highly branched structure composed of poly-glucose short chains of approximately 20–24 α D glucosyl residues connected by α1, 4 linkages with 5% αD-1, 6 glycosidic linkages.

Within starch granule the starch is packed into the so called A- or B-crystal patterns. The A-type starch has more densely packed structures and the B- type starch possesses shorter chain lengths Preiss, J. 2009 (Hizukuri, S., 1986, Imberty, A., 1988).

A-chains do not carry other chains, the B-chains which carry other chains through (1→6)-branches and each molecule contain a single C-chain, which carries the reducing end group (Peat, S., 1955).

The amylopectin is found as a complex level of molecular association where external segments of chains form clusters and the external segments of the chains form double helices that contribute to the crystalline structure. A cluster was defined by Bertoft, E., 2007 as a group of chains in which the internal chain length between the branches is less than 9 glucosyl residues (Laohaphatanaleart, K., 2010).

In the particular case of cassava, the amylopectin presents clusters of uniform size, DP 104–129 with 13–16 chains connected by B-chains, with internal chain length between branches of adjacent clusters of 10.5– 12.3 residues and no branches exist outside the clusters. (Laohaphatanaleart, K., 2010).

The physicochemical properties of the starch depend on its amylose: amylopectin ratio that inturn depends on the botanical source (Krisman, C., 1992).

Oustanding the content of one of these components is responsible for the architecture of starch granules and has an important influence on the physicochemical properties of starch and in consequence on their technological application.

THE CASSAVA STARCH

Isolation and Purification

The starch is isolated from cassava root *Manihot esculenta Crantz* (Euphorbiaceae) from Argentina. First, the skin is removed and secondly the root is cut in pieces which are homogenized in a blender during 4 hours (100g/ 100 ml), at 4°C with buffer glycine pH 6 soaked in 0.005 M NaHSO₃ to inhibit enzyme activity .The solution is filtered through muslin to separate the supernatant and the fibers. The fibers are submitted to six washes in the same conditions. The filtrate is collected to obtain starch by sedimentation.

The starch is washed with 0.1 N NaCl; twice with 96% ethanol: water mixtures (1:3), (2:3) and (3:1) successively. Finally, it is washed with 96% ethanol, acetone and ether, then dried and weighed.

To obtain the starch content, the total sugar content of polysaccharide is determined by phenol sulfuric acid method (Dubois, M., 1956) and later calculations considering:

Starch %= (A x F x 1000 x 1/1000 x 100/ w x 162/ 180)

A: absorbance against the blank absorbance
F: 100 ug of glucose/ absorbance for 100 ug of glucose
1000: volume correction factor
1/ 1000: conversion to milligrams
100/w: percentage sample weight
162/ 180: conversion free glucose to anhydre glucose

Molecular Fractionation

To evaluate cassava starch components the use of two basic methods allows to find out which has the best efficiency.

Method Based on the Differential Solubility of Starch Components in a Water- Butanol Saturated Solution

Is described by Schoch, T., 1956, modified by Curá, J., 1990 and later by Corcuera, V. 2007.

10 g of sample are suspended in four volumes of 3 % $HgCl_2$ pH 7 at 28 °C and stirred for an hour. The suspension is centrifuged at 5.000 x g for 15 min and the sediment is resuspended in 1-butanol: water (1: 7), autoclaved for 3 hours at 1 atmosphere, 110°C. After centrifugation at 3.000 x g for 20 min. two fractions are obtained: a) the amylopectin fraction found in the supernatant liquor that is removed after the addition of three volumes of 96 % ethanol, and b) the sediment containing the amylose fraction. Both fractions are then washed with ethanol, acetone and ether; dried and weighed successively for their later quantification and characterization.

The fractions are purified (by triplicate) by means of Biogel P6 chromatography (100 - 200 mesh), Amersham Biotech (100- 200 mesh) as was described in Kober, E., 2007 and then controlled on their reducing sugars content determined by Somogyi Nelson method (Aswell, G., 1966), total sugar content by phenol sulfuric acid method (Dubois, M., 1956), amylose or amylopectin content in the presence of the iodine reagent, a saturated $CaCl_2$ solution 800 g L^{-1}, containing I_2, 0.1 g L^{-1} and KI 1 g L^{-1} (Krisman, 1962) in UV-VIS Shimadzu spectrophotometer.

The ratio of total sugar to reducing sugar, (TS.RS-[1]), expresses the amount of total glucose units versus the amount of reducing ends of glucoses chains or free glucoses, both expressed as mMoles of glucose. mL^{-1}, not involving the concept of degree of branching chain.

Method Based on Fractionation of Starch with Concanavalin A

From the vegetal source *"Concanavalina ensiformis"* that has a high affinity with the carbohydrates, is proposed by Matheson, N., 1988, Colona, P., 1985, Yun, S., 1993 and Matheson, N., 1996. The starch sample is solubilized in 1.0 mL DMSO to pH 6, neutralized and incubated according to the methodology.

A schematic overview of the method is shown in Figure1. The starch solution mixed with Con. A in a concentration of 6.0 mg of protein/ mL gives origin to Con. A soluble fraction (supernatant) that is treated with sodium EDTA, the protein denatured and later discoulored and chromatographed on a Sepharose CL- 2B.

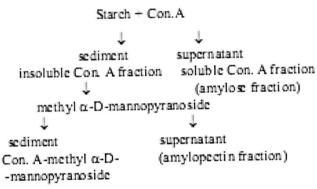

Figure 1.

The λ_{max} of the coloured complex is analyzed with iodine reagent (Krisman, 1962) in each α -D glucanes fraction collected.

It is considered that α1, 4 α1, 6 glucopolysaccharides in presence of iodine reagent (I_2 - KI in $CaCl_2$ saturated solution) present an absorbance between 380- 800 nm.

It is possible to estimate the length of the external chains of the α-D-glucan using the parameter "A", that is, the quotient between the maximum absorbance value of the polysaccharide and the maximum absorbance of the shoulder. The "A" value ranked from 1.17 to 1.30, when the λ_{max} ranges from 460 to 485 nm (Tolmasky, D., 1987, Krisman, C., 1962, Corcuera, V. 2007). On the other hand, when λ_{max} ranges from 608 to 632 nm, "A" values vary from 2.93 to 3.33, suggesting the presence of long and scarcely branched chains of amylose.

Table 1. Amylose and amylopectin of cassava *Manihot esculenta Crantz*

| | amylose | | amylopectin | |
Manihot esculenta Crantz	A	B	A	B
λ_{max}	588	586	488	485
*Glucopolysaccharide %	22	23	78	77
Iodine binding capacity %	19.4	19.4	1.01	1.02
Parameter "A"	2.3	2.3	1.20	1.20

A: method based on complexing the starch with lectin Con. A.
B: method based on differential solubility of the starch in butanol-water.
*: averages followed by the same letter do not differ significantly at $p \geq 0.01$
Iodine binding capacity %(IBC): mg iodine bound/ 100 mg polysaccharide
Parameter "A": 2.93 to 3.33 in linear chains, 1.17 to 1.30 in branched chains.

The amylose content is 22-23 % and amylopectin 77-78% not showing significative difference by either method.

The λ_{max} values for amylose and amylopectin are shown in Table 1. From this Table arises that the amylose λ_{max} values estimated by the method based on complexing the starch with lectin Con. A or the method based on differential solubility of the starch in butanol-water is 588 nm and 586 nm. For the average amylopectin λ_{max} values obtained 488 and 485 respectively, so that only a difference of 0.15% is found between averages achieved by both methodologies (Figure 2 A-B).

In the case of amylopectin cassava starch a brown- violet coloration and a λ max of 488 nm (linear regression quotient R= 0.9908) denotes the presence of branched chains with the characteristic "A" value 1.20. In the case of amylose a blue coloured complex and λ max of 588 nm(linear regression quotient R= 0.8999).The "A" value 2.3 indicates the presence of chains that are not completely lineal, but have a low degree of ramification and this must be proved by other methodologies.

One practice criteria considered for the determination of iodine binding capacity for amylose, amylopectin and starch is based on a spectrophotometric method (Karve, M., 1992) using a constant volume titration, method that is in agreement with the conventional potentiometric iodine titration methods (Banks, W., 1974).Table 1 shown IBC for amylose and amylopectin19.4 and 1.01 respectively.

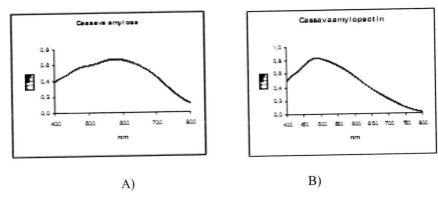

A)   B)

Figure 2. A-B: Absorption spectrum of amylose and amylopectin.

STARCH MODIFICATION

Biodegradable Plastic Material

To form a matrix of a new biodegradable plastic material by compression molding, cassava starch is complemented with proteins (Salmoral, E., 2000). The conformed plastic material is defined as biodegradable since it can be consumed by microorganisms and reduced to simple compounds (Floccari, M., 2001).The proteins as well as the process by compression molding to produce the plastic material are previously described (Salmoral, E., 2000).

One of the undesirable characteristics of this plastic is its high water absorption capacity, a problem solved in our laboratory by irradiation of starches previous to their incorporation to the plastic blend.

The cassava starch irradiation is carried out with gamma rays from a Co-60 source, applied doses of 10, 25 and 50 kGy, at a dose rate of 5.56 Gy.s^{-1} in Semi Industrial Irradiation Plant of Ezeiza Atomic Center, Buenos Aires, Argentina. The dry starches are packed in closed polyethylene flasks, under temperature, in the irradiation chamber, never over 30°C, and dichromate dosimeters for high doses are used as was expressed previously (González, M.E., 2002).

The degradation of starch by influence of ionizing radiation is well known (Adam, S., 1983; Raffi, J., 1981; Sokhey, A., 1993; Mac Arthur, L., 1984; Michel, J., 1980), Kober, E., 2007). Amylose and amylopectin diminish their degree of polymerization as a consequence of γ radiation, as was shown by the increase of reducing sugars when the radiation dose increases from 0 to 30 kGy in native waxy and manioc (Raffi, J., 1981).

The changes produced in the molecular starch cassava structure as a consequence of gamma irradiation and analyzed by means of Biogel P6 chromatography (100 - 200 mesh), in a 10 cm length and 0.8 cm diameter column in 0.1 M, pH 6 buffer acetate –pyridine are shown in Table 2. For these studies 10 mg of starch are solubilized in 0.1 M NaOH, neutralized to pH 6.0 with HCl 0.1 M and applied to the column.

Table 2. Biogel P-6 Chromatography of cassava starch submitted to γ radiation

chromatography		cassava starch			
peak	fraction	radiation dose (kGy)			
		0	10	25	50
1	3-5 = 3ml	97	58.10	46.50	40.00
2	9 = 1ml	0	15.46	10.13	18.00
3	15 = 1ml	0	14.50	18.20	20.00
4	19 = 1ml	1	8.3	15.49	12.00
5	25 = 1ml	2	3.5	9.92	10.00

Fractions of 1 mL are collected and controlled: 1) reducing sugars content (RS), determined by Somogyi Nelson method (Aswell, G., 1966); 2) total sugar content (TS) by phenol sulfuric acid method (Dubois, M., 1956) and 3) glucopolysaccharides in the presence of iodine reagent (Krisman, C.R., 1962).

In the first peak, we can appreciate that the presence of α glucanes are in gradual degradation, from 0 to 50 kGy. This peak corresponds to chains of more than 40 glucose units, according to the paper chromatogram shown in Table 3.

This allows us to deduce the extent of depolymerization of starch during irradiation treatment. The effects of gamma radiation on the plastic materials in Figure 3 show that plastic prepared with non-modified starch absorbs 65% and gradually reduces its capacity by 60% reaching a final value of 28% when the dose applied to the starch increases until 50 kGy. A 10-kGy radiation dose applied is enough to produce a reduction of the water absorption capacity of the plastic of 60%.

This behavior might be explained as a consequence of radiolytic products being produced in the case of the amylopectin rich cassava starch, thus generating more branching points in the starch-protein matrix.

The ability of molded products to absorb water is measured according to ASTM D- 570, (1995). The samples are conditioned in an oven for 24 h at 50 ± 3 °C, then cooled in a desiccator and weighed to the nearest 0.001g. The samples are immersed in distilled water for 2 h at 23 °C and immediately weighed to determine the absorbed water. Differences between means were considered significant when $p \leq 0.05$ (Student test).

Table 3. Quantification of glucosa

Chromatography Biogel P6 fraction	Glucose numbers
1	$> 40 \pm 0.05$
2	$8\text{-}12 \pm 0.09$
3	$5\text{-}8 \pm 0.1$
4	$3\text{-}4 \pm 0.09$
5	$2\text{-}3 \pm 0.09$

The ratio of total sugar (TS) to reducing sugar, (TS.RS-1), expresses the amount of total glucose units versus the amount of reducing ends of glucoses chains or free glucoses, both expressed as mMoles of glucose.mL^{-1}. TS.RS-1 values obtained for the first peak express the degree of polymerization. The elution volume of each chromatographic peak is considered to calculate percentage values in order to facilitate the comparative study.

It is known that starches are not used as emulsifying agents because of their chemical characteristics, unless their hydroxyl groups are modified by chemical reactions. Studies over the last years proposed modification of cassava starch (a, b-Martin, A.M., 2010), and the effect of emulsifiers on fermented cassava starches (Nunfor, F., 1996).

This chapter reports studies about water in-oil micro-emulsion widely used in cosmetic and pharmaceutical products. Branched α1-4, α1-6 D-glucopolysaccharide from cassava are chemically modified by esterification with fatty acid chloride to obtain a non-polar agent with emulsifying capacity.

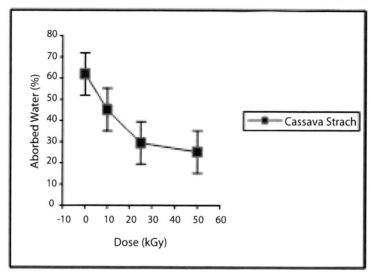

Figure 3. Water absorption capacity of plastic material.

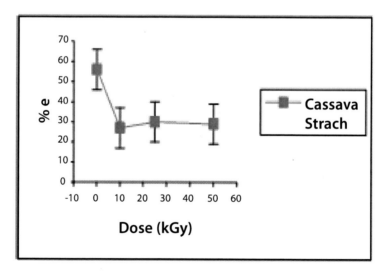

Figure 4. Mechanical property.

Microemulsion with Esterified Starch

The isolated and purified glucopolysaccharide from *Manihot esculenta Crantz* is chemically modified by esterification reaction with lauroyl chloride to obtain a non-polar ester. The esterification reaction is performed according to Aburto, J., 1990, and Roberts, H., 1967 methods. The carbohydrate esterification utilized formic acid (0.2 - 20 eq /glucose) and 4-10 eq /glucose lauroyl chloride. Emulsions are prepared as the basic model with preserved water, mineral oil, olive oil, sorbitan olivate, and glucopolysaccharide ester, stirring at 100 rpm, 80°C until the mixture was emulsified using Brawn homogenization equipment (100 rpm).

Samples for FT-IR analysis are prepared as pellets in BrK and measured in the range of 600 - 4600 cm-1 in a Nicolet FT-IR spectrophotometer.

If theoretical considerations about the maximum DS of α1-4, α1-6 D-glucan ester indicate the same number of hydroxyl groups per D-glucopyranosyl unit, DS = 3. Experimentally a high DS = 2, both glucopolysaccharide fractions studied presented DS = 1 .A change on their colour from white to light- brown is detected in the special case of high substitution during DS process.

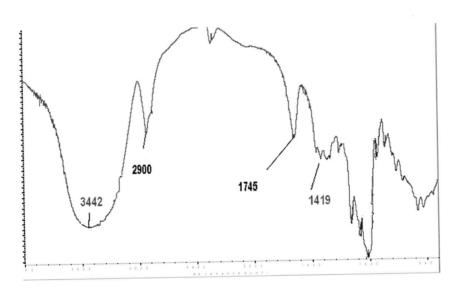

Figure 5A. FT-IR spectra of non-esterified cassava starch.

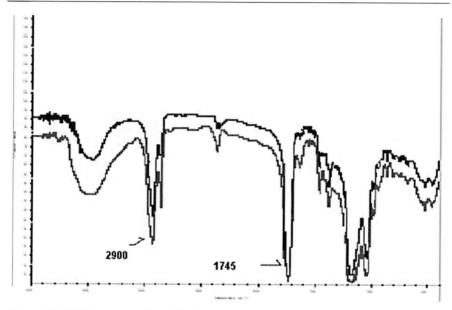

Figure 5B. FT-IR spectra of esterified cassava starch.

In Figure 5A and Figure5 B the esterification is shown by the presence of an intense carbonyl stretching absorption at approximately 1745 cm-1; different from the lauroyl absorption at 1800 cm-1 and the increase in the intensity of the CH_2 stretching at 2940 cm-1 in relation to the OH absorption in the 3500 cm-1 region.

The esterified starch obtained with different equivalent/glucose (1) 4 equiv. (superior); (2) 6 equiv. is shown in Figure 5B. The degree of substitution obtained is Ds = 0.95 and DS = 1 respectively (Figure 5B).

The rheological properties are an important factor in the stability of the emulsions. When the esterified starch is tested as self-emulsifying in water-in-oil emulsions the viscosity is increased

So they are used as rheological additive to increase the viscosity. It was found that the viscosity increased in the emulsions having starch ester, as can be seen fin Table 4.

Table 4. Viscosity of the emulsions with and without the esterified polysaccharide (Accuracy 2.0%)

vegetal source of α D glucan	viscosity x 10^{-3} (cP)	
	Control emulsion	Additivated emulsion
Cassava	130	180

CONCLUSION

Cassava starch molecular fractionation may be appropriately and successfully done by the method based on the starch differential solubility in butanol saturated water or by the carbohydrate-protein complex lectin Concanavalin A method. Similar amylose and amylopectin content as starch qualitative composition is found.

The modification of starch by chemical reactions altering the OH groups or by gamma radiation action provide a possibility to have industrial application. The effect of the incorporation to the plastic blend of cassava starch previously irradiated, modified the water absorption capacity of the resulting plastic composite. Water absorption of the resultant product is reduced by 60%. Their elongation declines by 45% but it does not affect other properties of the plastic material, which is very useful for special cases.

Cassava starch may play the role of network-forming that yields non-covalent molecular associations incorporated in micro-emulsions after being esterified with formyl- laureate chloride, as demonstrated by FT-IR. It produces a 30% increase in viscosity with a positive influence on the stability.

REFERENCES

Aburto, J.; Alric, I., Borredon, E. Preparation of long-chain esters of starch using acid chlorides in the absence of an organic solvent. *Starch/ Stärke* (1999) 51: 132-135.

Adam, S. Recent developments in radiation chemistry of carbohydrates. In: Recent Advances in Food Irradiation. Elias, P.S. and Cohen, A.J. (Eds.), Elsevier Biomedical Press, Netherlands, pp. 149- 170, 1983.

Ashwell, G. New colorimetric methods of sugar analysis. In: Complex Carbohydrates, Methods in Enzymology vol. VIII. E.F. Neufeld, V. Ginsburg, S.P. Colowick and N.O. Kaplan, Editors, Academic Press, NY pp. 85–95 (1966).

Banks, W.; Greenwood, C.; Muir, D. The characterization of starch and its components. Part 6 A critical comparison of the estimation of amylose content by colorimetric determination and potentiometric titration of the iodine complex. *Starch/ Stärke* (1974) 26: 73- 78.

Bertoft, E. Carbohydr. Polym. 70 (2007) 123–136.In: Biochemistry and Molecular Biology of Starch Biosynthesis. Chapter 4, J. Be Miller and R.

Whistler (Eds.) Starch: Chemistry and Technology, 3rd edition, 2009. Academic Press, NY, USA. ISBN 978-0-12-746275-2.

Colonna, P.; Biton, V.; Mercier, C. Interactions of concanavalin A with α-D glucans. *Carbohydrate Research* (1985) 137: 151-166.

Corcuera, V.; Salmoral, E.M.; Salerno, J.C.; Krisman, C.R. Starch molecular fractionation of bread wheat varieties. *Agriscientia (2007)* XXIV (1):11-18.

Curá, J. A.; Krisman, C. R. Cereal grains: A study of their (α1, 4) - (α1, 6) glucopolysaccharides composition. *Starch/ Stärke* (1990) 42: 71- 175.

Cúra, J. A.; Per-Erik, J.; Krisman, C. R. Amylose is not strictly linear. *Starch/ Stärke* (1995) 47: 207- 209.

Dubois, M.; Gilles, K.; Hamilton, J.; Rebers, P.; Smith, F. Colorimetric method based on phenol sulfuric acid. *Analytical Chemistry* (1956) 28: 356- 359.

Floccari, M.E.; Traverso, K.; González, M. E.; Salmoral, E. M. Biodegradability studies of irradiated plastic material based on protein-starch matrix. Proceedings Naro.Tech- Messe und Kongress, Erfurt, Alemania (2001) 314: 1- 5.

González, M. E.; Salmoral, E. M.; Floccari, M.; Traverso, K. Effects of gamma radiation on a plastic material based on bean protein. International Journal of Polymeric Materials, Gordon and Breach Publishing. NY. (2002) 51 (8): 721-731.

Hizukuri, S. Polymodal distribution of the chain length of amylopectin and the crystalline structure of starch granule. *Carbohydrate Research* (1986) 147: 342- 347.

Imberty, A.; Chanzy, H.; Perez, S.; Buléon, A. The double helical nature of the crystalline part of A starch. *Journal of Molecular Biology* (1988) 20: 365-378.

Karve, M. S. and Kale, N. R. A spectrophotometric method for the determination of iodine binding capacity for starch and its components. *Starch/ Stärke* (1992) 44 (1 S): 19-21.

Kober, E.; González, M.E.; Gavioli, N.; Salmoral, E.M. Modification of water absorption capacity of a plastic based on bean protein using gamma irradiated starches as additives. Radiation Physics and Chemistry Elsevier, Irlanda (2007) 76 (1): 55-60.

Krisman, C.R. A method for the colorimetric estimation of glycogen with iodine, *Anal. Biochem.* (1962) 4(1): 17–23.

Krisman, C.; Curá, A.; Salmoral, E.M. Glucopolysaccharides α,1-4 α,1-6 contained in crops of agronomic interest. *Starch/ Stärke* Netherlands (1992) 44:6-7.

Laohaphatanaleart, K.; Piyachomkwan, K.; Sriroth, K.; Bertoft, E. The fine structure of cassava starch amylopectin. Part 1: Organization of clusters. *International Journal of Biological Macromolecules* (2010) 47: 317– 324.

Larson, B.L.; Gilles, H.A.; Jennes, R. Amperometric method for determining the sorption of iodine by starch. *Analytical Chemistry* (1953) 25(5): 802– 804.

Mac Arthur, L.A.; D´applonia, B.L. Gamma radiation of wheat II. Effects of low- dosage radiations on starch properties. *Cereal Chemistry* (1984) 61: 321- 326.

Manners, D. J. Recent developments in our understanding of amylopectin structure. *Carbohydrate Polymer* (1989) 11: 87– 112.

Matheson, N.K.; Welsh, L.A. Estimation and fractionation of the essentially unbranched amylose and branched amylopectin component of starches with concanavalin A. *Carbohydrate Research* (1988) 180: 301-313.

Matheson, N.K. The chemical structure of amylose and amylopectin fractions of starch tobacco leaves during development and diurnally nocturnally. *Carbohydrate Research* (1996) 282 (2): 247-262.

a-Martín, A. M.; González, M.E.; Etienot,C.; Gavioli, N.; Salmoral, E. M.Laureate starch ester as additive in water-in-oil emulsions. XXVI Congress of the International Federation of Societies of Cosmetic Chemists. Buenos Aires, Argentina, September, 2010.Number 0173.

b- Martín, A.M.; González, M. E.; Etienot, C.; Urreaga, G.; Salmoral, E. M. Comparative esterified glucopolysaccharides in water-in-oil microemulsions. XXVI Congress of the International Federation of Societies of Cosmetic Chemists. Buenos Aires, Argentina, September, 2010 Number 0391.

Michel, J. P.; Raffi, J.; Saint- Lebe, L. Experimental study of the radiodepolymerization of starch. *Starch* (1980) 32: 295- 298.

Morrison, W. R.; Laignelet, B. An improved colorimetric procedure for determining apparent and total amylose in cereal and other starches. *Journal of Cereal Science* (1983) 1: 9– 20.

Nunfor, F.; Walter, W.; Schwartz, S. Effect of emulsifiers on the physical properties of native and fermented cassava starches. *J. Agric. Food Chem.* (1996) 44: 2595- 2599.

Peat, S.; Whelan, W.J.; Rees, W.R.D. Enzyme: a disproportionating in potato juice. Nature 172: 158-160, *J. Chem. Soc., Chem. Commun.* (1952): 4546–4548.

Peat, S., Whelan, W.J., Edwards, T.E. Journal Chem. Soc. (1955):355-359.

Preiss, Jack. Biochemistry and molecular biology of starch biosynthesis, Chapter 4 in Starch: *Chemistry and Technology*, 3rd edition, pg. 83-148, J. Be Miller and R. Whistler (Eds.).Academic Press, NY, USA. ISBN 978-0-12-746275-2 (2009).

Raffi, J. J.; Agnel, J. P.; Thieri, C .J.; Fréjaville, C. M.; Saint-Lèbe, R. L. Study of γ irradiated starches derived from different foodstuffs: A way for extrapolating whole someness data. *J. Agric. Food. Chem.* (1981) 29: 1227- 1232.

Roberts H. Starch Derivatives in: Starch Chemistry and Technology. Whistler et al (Eds.), Acad. Press, NY, 1967.

Robin, J. P.; Mercier, C.; Charbonniere, R.; Guilbot, A. Lintnerized starches. Gel filtration and enzymatic studies of insoluble residue from prolonged acid treatment of potato starch. *Cereal Chem.* (1974) 51: 389- 406.

Roger, P.; Tran, V.; Lesec, J.; Colonna, P. Isolation and characterization of single chain amylose. *Journal of Cereal Science* (1996) 24: 247– 262.

Salmoral, E. M.; González, M. E.; Mariscal, P. Biodegradable plastics made from bean products. Industrial Crops and Products, Elsevier *Science* (2000) 11 (2-3): 217-225.

Schoch, T. J; Maywald, E. C. Fractionation of starch by selective precipitation with butanol. *Analytical Chemistry* (1956) 28: 382- 389.

Sokhey, A. S.; Chinnaswamy, R. Chemical and molecular properties of irradiated starch extrudates. *Cereal Chemistry* (1993) 70: 260- 268.

Takeda, Y.; Shitaozono, T.; Hizukuri, S. Molecular structure of corn starch. *Starch* (1988) 40, 2S: 51-54.

Tolmasky, D. S.; Krisman, C. R. The degree of branching in (α1, 4) (α1, 6) linked glucopolysaccharides is dependent on intrinsic properties of the branching enzymes. *European Journal of Biochemistry* (1987) 168: 393-397.

Yun, S. H.; Matheson, N. K. Structures of the amylopectins of waxy, normal amylose extender and wx: ae genotypes and of phytoglycogen of maize. *Carbohydrate Research* (1993)243(2): 307-321.

INDEX

food, 5, 7, 17, 19, 22, 23, 25, 33, 50, 51, 55,
    57, 85, 86, 87, 90, 91, 92, 93, 97, 99,
    100, 101, 102, 103, 110, 114, 115, 116,
    117, 118, 119, 123, 125, 147, 148, 149,
    150, 153, 154, 156, 157, 161, 162, 167,
    169, 171, 174, 182
formation, 5, 9, 10, 34, 44, 96, 105, 124,
    138
foundations, 159
friction, 125, 126
frost, 62
fructose, 1, 2, 23, 25, 28, 39, 45, 88, 91
functional food, 157
fungal infection, 63
fungi, 19, 43, 87, 97
fungus, 22, 25, 37, 41, 86, 96
fusion, 100, 140

G

gait, 167, 168, 173
gamma radiation, 182, 188, 193, 194
gamma rays, 187
garbage, 92
gel, 13, 20, 105, 119
gelatinization temperature, 9, 104
gelation, 105
gene pool, 154
genes, 150, 154, 156, 157
genetic diversity, 151, 160, 161
genetic marker, 160
genotype, 57, 66, 73
genus, 146
geometry, 129
glass transition, 139
global climate change, 150
global warming, 159
glucoamylase, 10, 16, 24
glucose, 1, 2, 7, 9, 10, 14, 16, 21, 24, 25, 28,
    38, 39, 40, 45, 85, 88, 91, 92, 115, 137,
    139, 181, 182, 184, 185, 188, 189, 191,
    192
glucoside, 160, 161
glutamate, 3, 87
glutamic acid, 23, 28, 43, 48, 87, 96

glycerol, 11, 104, 108, 109, 110, 122, 123,
    124, 125, 136, 140, 141
glycine, 183
glycogen, 194
glycol, 19, 31, 124
glycoside, 149, 160, 166, 167, 171
gracilis, 42
granules, 3, 9, 16, 85, 100, 137, 138, 140,
    182, 183
greenhouse, 58
greening, 30
groundwater, 82
growth, 4, 5, 12, 14, 19, 21, 23, 37, 38, 39,
    40, 43, 44, 45, 53, 56, 58, 59, 61, 62, 64,
    66, 68, 69, 72, 73, 75, 76, 84, 86, 88, 90,
    93, 105, 110, 112, 113, 115, 121, 160,
    182
growth rate, 21, 113
gums, 100, 101

H

harbors, 150
hardness, 103
harvesting, 3, 57, 147, 157
health, 149, 154
height, 56, 59, 60, 61, 62, 63, 65, 66, 68
high-value market compounds, viii, 33
history, 105, 153, 169
House, 30
human, 100, 122, 146, 147, 154, 161, 171
human body, 171
human health, 154
humidity, 4, 57, 61
hydrogen, 80, 90, 97, 101, 123, 124, 166,
    167, 172
hydrogen bonds, 123, 124
hydrogen cyanide, 167, 172
hydrogen gas, 80, 97
hydrogenation, 17
hydrolysis, 9, 10, 13, 14, 15, 24, 27, 29, 30,
    31, 38, 39, 44, 83, 84, 85, 86, 88, 94, 95,
    167
hydrophilicity, 120
hydroxide, 17, 42

monosodium glutamate, 87
Moon, 27, 98
morphology, 32
mortality, 57
mosaic, 151, 160
motor neuron disease, 163, 168, 176
moulding, 122, 125, 126, 127, 128, 129,
   138, 141
Mozambique, 168, 176, 177, 180
multilayer films, 103
multiplication, 55, 57, 58, 60, 62, 64, 65, 66,
   73, 74, 77
municipal solid waste, 97
mutation, 12
mycelium, 88

N

NaCl, 22, 113, 184
nanocomposites, 30
national product, 101
natural food, 87
nematode, 45
Nepal, 179
nerve, 168, 173
Netherlands, 142, 193, 195
neurodegeneration, 171
neurological disease, 177
neurological disorders, 149, 165, 166, 167
neuropathy, 165, 167, 168, 173, 174, 175,
   176, 178, 179
neurotoxicity, 178
niacin, 167
nicotinamide, 174
Nigeria, 31, 51, 152, 167, 169, 172, 174,
   176, 179, 180
nitrogen, 14, 15, 33, 34, 40, 42, 44, 82, 83,
   87, 90, 91, 92, 97, 154
nitrogen fixation, 155
nodes, 56, 60, 61, 62, 63, 65, 67, 68, 70, 147
non-polar, 123, 189, 191
normal distribution, 132, 134
North America, 101
nuisance, 82

nutrient, 15, 22, 35, 56, 57, 60, 61, 67, 68,
   69, 70, 72, 73, 74, 75, 94
nutrients, 5, 17, 21, 33, 36, 38, 40, 65, 73,
   74, 89, 90, 100, 102, 147, 148, 154, 179

O

OH, 124, 126, 192, 193
oil, 49, 88, 93, 96, 102, 155, 163, 189, 191,
   192, 195
olive oil, 191
opacity, 103
opisthotonus, 175
opportunities, 79, 118, 147
optic nerve, 173
optimization, 31, 53, 128, 134
organ, 147
organic acids, 2, 26, 32, 80, 86
organic matter, 35
organism, 14, 21
organs, 71, 100
osmosis, 119
oxidation, 87
oxygen, 25, 34, 35

P

pallor, 168
paralysis, 149, 173, 179
Pareto, 132, 134, 135
partition, 71
pathogenesis, 178, 180
pathogens, 17, 57, 149
pathways, 155, 156, 168
peat, viii, 56, 60, 69
permeability, 19, 107, 108, 109, 120
Peru, 27
pests, 46, 55, 57, 146, 151
pH, 10, 11, 15, 16, 17, 21, 23, 36, 37, 59,
   85, 88, 102, 110, 111, 115, 117, 150,
   183, 184, 185, 188
pharmaceutical, 17, 25, 86, 87, 92, 93, 157,
   189
phase transitions, 127
phenol, 184, 185, 188, 194

Q

R

T

U

V